The Demand for Responsiveness in Past U.S. Military Operations

STACIE L. PETTYJOHN

Prepared for the Department of the Air Force
Approved for public release; distribution unlimited

RAND PROJECT AIR FORCE

For more information on this publication, visit www.rand.org/t/RR4280

Library of Congress Cataloging-in-Publication Data is available for this publication.
ISBN: 978-1-9774-0657-6

Published by the RAND Corporation, Santa Monica, Calif.
© 2021 RAND Corporation
RAND® is a registered trademark.

Cover: U.S. Air Force/Airman 1st Class Gerald R. Willis.

Support RAND
Make a tax-deductible charitable contribution at
www.rand.org/giving/contribute

www.rand.org

Preface

The Department of Defense (DoD) is entering a period of great power competition at the same time that it is facing a difficult budget environment. The joint force remains unprepared for a great power war because for the past two decades it has been focused on wartime counterinsurgency and counterterrorism operations, while remaining committed to expensive legacy modernization programs that do not solve the operational challenges that a great power adversary would present. The combination of high and continuing operational demands with reduced resources has greatly stressed the joint force as near-term readiness increasingly competes for funds with programs needed to meet future challenges.

To address these and related policy issues, this study identified four questions for analysis: (1) What has been the historic demand for joint force responsiveness over the last sixty years? (2) What are the responsiveness demands associated with current U.S. strategy, operational plans, and planning scenarios? (3) What lessons can be learned from the New Look Era when airpower was widely viewed as the cornerstone of U.S. national security? (4) How have U.S. Air Force public narratives and popular attitudes toward airpower evolved over the last century? This report presents findings from research exploring historical demands for responsiveness. Other forthcoming reports address the other three questions.

The research reported here was commissioned by U.S. Air Force (USAF) Quadrennial Defense Review Office (HAF/CVAR) and conducted within the Strategy and Doctrine Program of RAND Project AIR FORCE.

RAND Project AIR FORCE

RAND Project AIR FORCE (PAF), a division of the RAND Corporation, is the USAF's federally funded research and development center for studies and analyses. PAF provides the USAF with independent analyses of policy alternatives affecting the development, employment, combat readiness, and support of current and future air, space, and cyber forces. Research is conducted in four programs: Force Modernization and Employment; Manpower, Personnel, and Training; Resource Management; and Strategy and Doctrine. The research reported here was prepared under contract FA7014-16-D-1000.

Additional information about PAF is available on our website: http://www.rand.org/paf.

This report documents work originally shared with the USAF January 2014. The draft report, issued December 2014, was reviewed by formal peer reviewers and USAF subject-matter experts.

Contents

Figures

Tables

Summary

The 2018 National Defense Strategy made "rebuilding readiness" a priority as a step toward "build[ing] a more lethal Joint force," and the Trump administration has used low levels of readiness to justify increased defense spending in the FY 2019 budget.[1] This raises two questions: What is readiness, and why is it important? At the most basic level, readiness means that forces are able to carry out their designated missions and responsibilities.[2] Most importantly, when they are called on, ready forces are able to quickly mobilize, deploy, and execute an operation.

Because the relatively high level of readiness across the Department of Defense (DoD) is its single greatest expense, there are those that question whether readiness needs to be improved or whether the specter of a hollow force is one that is raised to justify increased defense expenditures. One counter to these arguments is that readiness is a critical factor that enables the U.S. military to be highly responsive, which is something that is expected of it. This report aimed to contribute to this debate on readiness by exploring the requirement for responsiveness across time.

This historical analysis has several components. First, I examine ten cases to determine the level of responsiveness and what role, if any, responsiveness played in achieving the U.S. military and political objectives. These cases, which are listed in Table 1.1, were chosen because they varied on several key dimensions, including the type and location of the operation, the date on which they occurred, and the level of responsiveness.[3] The cases, therefore, cover a broad range of past demands and present differing outcomes. Because this work was done for the U.S. Air Force (USAF), particular attention is paid to the role of airpower in each of the cases. Yet the cases tell the story of how quickly the joint force deployed. The responsiveness of different types of forces could not be examined in isolation from the entire operation because, at least in part, the forces deployed were affected by the pool of available forces across domains. So while airpower is at times called out in the narrative, the study is examining the demand for responsiveness within the joint force—not just USAF.

Second, I created a database of U.S. military operations from 1945 through 2013 and identified how often the joint force has been called on to respond to multiple contingencies

---

[1] DoD, *Summary of the 2018 National Defense Strategy of the United States of America: Sharpening the U.S. Military's Competitive Edge*, Washington, D.C., 2017, p. 5.

[2] Todd Harrison, "Rethinking Readiness," *Strategic Studies Quarterly*, Vol. 8, No. 3, Fall 2014, p. 38. For more on conceptual confusion over what readiness is, see Richard K. Betts, *Military Readiness: Concepts, Choices, Consequences*, Washington, D.C.: Brookings Institution, 1995, pp. 35–43; Russell Rumbaugh, *Defining Readiness: Background and Issues for Congress*, Washington, D.C.: Congressional Research Service, June 14, 2017. In this report, I use what Rumbaugh calls the more narrow definition of operational readiness.

[3] This excludes other critical capabilities that USAF provides, including space and cyber, which did not lend themselves to an open-source analysis.

simultaneously. This was done because it was clear that looking at the size of the force and urgency associated with individual cases understated the potential total responsiveness required because operations can take place at the same time.

This historical analysis demonstrates that U.S. forces have been called on to rapidly respond to crises many times since 1950 and that responsiveness was important—to varying degrees—to achieving the United States' political aims in all but one of the cases examined. Quick reactions are often needed in response to unanticipated or out-of-the blue events, such as terrorist attacks and hijackings, and they are also important when the United States is seeking to deter aggression, prevent an opponent from rapidly achieving its objectives, or assist a partner or ally that is under attack. Conversely, responsiveness tended to be less important when critical U.S. interests were not at stake but decisionmakers chose to use force as a coercive tool.

Overall, in these cases U.S. forces have proven quite responsive and typically had enough military capability in place to carry out the operation within weeks of the deployment order being issued. It was only in the instances of deterring or intervening in a major war where the U.S. required a large joint force, including heavy ground troops, that it took months to mass the requisite forces.

In the ten cases, there were many recurring factors that facilitated rapid responses, including having forward-based or deployed forces proximate to the area of operations, political decisiveness, prepared en route infrastructure, prepositioned equipment, agreed-upon base access, strategic airlift, and prepared contingency and deployment plans. Not surprisingly, the absence of these factors hindered responsiveness.

Yet the past is not always a good predictor of the future security environment. As we look to the future, what range of contingencies is the joint force likely to be called on to carry out? U.S. forces have been frequently called on to respond to multiple operations at the same time, but many of these were small-scale contingencies (SSCs). The Trump administration has prioritized strengthening deterrence against potential great power adversaries—in particular Russia and China—which requires a large joint force capable of rapidly responding and preventing aggressors from completing a fait accompli.

Going forward, the United States must consider the balance between readiness for smaller-scale crisis response operations and being prepared to deter and defeat a great power in major combat. During the Cold War, when the United States prioritized the latter, the joint force was still called on to conduct smaller operations, and simultaneity events were common. Since the end of the Cold War, the balance shifted in favor of SSCs, and the demand for simultaneous operations has increased significantly.

It is not clear that the U.S. military can continue to try to do everything. Preparing for great power competition is likely to require U.S. policymakers to be more judicious in when they task U.S. forces with responding to other situations. The strains on the force are particularly large today because the U.S. military is mainly based in the continental United States (CONUS). Additional forces are sourced through temporary rotations, which is a greater drain on service force structure compared to permanently forward-based forces. Even if U.S. forces engage in fewer SSCs, because simultaneity events happen regularly, DoD needs to consider the total possible demand for responsiveness.

Finally, how responsive is airpower? Generally, airpower has proven to be extremely responsive. In nearly all of the cases examined, 50 percent of the air forces needed were in place in three weeks or less. Moreover, unlike other types of forces, air forces offer global responsiveness. Because of their range and speed, ready air forces can quickly respond to events nearly anywhere in the world. They, therefore, do not need to be in position ahead of the time to respond within days or weeks to unforeseen events around the world. Their responsiveness, however, does depend on the availability of air bases with adequate infrastructure to support operations and adequate throughput to flow forces into a distant theater. While air forces are able to quickly respond to contingencies across the globe, it is not clear that they are always the right or preferred force for a given situation. For instance, there are questions about whether airpower alone could stop a Russian short-warning invasion of the Baltics, given Russia's dense network of air defenses and how rapidly the invading force could advance. There simply might not be enough time for the North Atlantic Treaty Organization's (NATO's) air forces to deliver enough weapons to stop an invading force of that size.[4] By themselves, air forces also have experienced difficulties conducting rescue operations as was evidenced in the *Mayaguez* rescue operation. So while airpower is highly responsive, it might not always be the capability that is required for a given situation.

This historical analysis cannot specify the exact level of readiness that the joint force or USAF should maintain. Because readiness exists along a spectrum, different readiness models and mixes of levels of readiness may be appropriate for different types of forces. But the analysis does suggest that the United States has demanded a relatively ready military since World War II and that if the American people and American policymakers continue to expect their military to be able to quickly respond to events anywhere in the world, maintaining a relatively high level of readiness will be necessary.

---

[4] David A. Shlapak, *The Russian Challenge*, Santa Monica, Calif.: RAND, PE-250-A, 2018, pp. 10–11.

Acknowledgments

I thank Lt Gen Steven Kwast and Maj Gen David Allvin for sponsoring this study and the project action officer Sam Szvetecz for his guidance during the study. I also owe a debt of gratitude to the Air Force Historical Research Agency and the Pacific Air Forces Command Historian's Office for providing historical materials. At RAND, Paula Thornhill, Alan Vick, Jeff Hagen, Karl Mueller, David Shlapak, and Lt Col Aaron Cowley all offered valuable advice on how to improve the research. Also, thanks to Robert Guffey for creating the simultaneity analysis figures. Finally, I thank Sean Zeigler and Michael O'Hanlon for their thorough and constructive comments that significantly improved the document.

Abbreviations

ACTORD	activation order
AO	area of operations
ARG	amphibious ready group
AWACS	Airborne Warning and Control System
CASF	composite air strike force
CONUS	continental United States
CVBG	carrier battle group
DoD	Department of Defense
DPRK	Democratic People's Republic of Korea
FEAF	Far East Air Forces
IDF	Israeli Defense Forces
JCS	Joint Chiefs of Staff
KSA	Kingdom of Saudi Arabia
MAC	Military Airlift Command
MEB	marine expeditionary brigade
MEU	marine expeditionary unit
NAC	North Atlantic Council
NATO	North Atlantic Treaty Organization
NKPA	North Korean People's Army
OAF	Operation Allied Force
PRC	People's Republic of China
ROK	Republic of Korea
SSC	small-scale contingencies
UAE	United Arab Emirates
UN	United Nations
USAF	U.S. Air Force
USMC	U.S. Marine Corps
USN	U.S. Navy

1. The Link Between Responsiveness and Readiness

The 2018 National Defense Strategy made "rebuilding readiness" a priority as a step toward "build[ing] a more lethal Joint force," and the Trump administration has used low levels of readiness to justify increased defense spending in the FY 2019 budget.[1] This raises two questions: What is readiness, and why is it important? At the most basic level, readiness means that forces are able to carry out their designated missions and responsibilities.[2] Readiness is not a binary concept where either the force or a unit is ready or not. For example, a fighter squadron that just returned from a deployment to Afghanistan may be highly proficient at air-to-ground operations but less capable of executing air-to-air operations, which are not required in that operation. In short, when they are called on, ready forces are able to quickly mobilize, deploy, and execute a given operation.

The current readiness deficit has multiple sources.[3] The U.S. military as a whole, and the U.S. Air Force (USAF) in particular, began to alert policymakers that their readiness was suffering as a result of the incessant demand for combat forces to fight in Afghanistan and Iraq, along with the increased demand for forward deployed forces to reassure and deter in Europe and Asia.[4] Additionally, U.S. forces were expected to be prepared to execute at a moment's notice unforeseen contingency operations, ranging from an air campaign to oust Libyan president Muammar Qaddafi in 2011 to disaster relief efforts in the Philippines in 2013 after Typhoon Haiyan. The problems created by sustained high demand for forces were exacerbated by the 2011 Budget Control Act and the implementation of sequestration in 2013. Secretary of Defense James Mattis has asserted that "no enemy in the field has done as much to harm the readiness of U.S. military than the combined impact of the Budget Control Act's defense spending caps" and the budget uncertainty created by Congress's reliance on continuing resolutions.[5]

---

[1] DoD, 2017, p. 5.

[2] Harrison, 2014, p. 38; For more on conceptual confusion over what readiness is, see Betts, 1995, pp. 35–43; Rumbaugh, 2017. In this report, I use what Rumbaugh calls the more narrow definition of operational readiness.

[3] Some have questioned whether there is a readiness crisis: David Petraeus and Michael O'Hanlon, "The Myth of a U.S. Military 'Readiness' Crisis," *Wall Street Journal*, August 9, 2016; Todd Harrison, "Trump's Bigger Military Won't Necessarily Make the US Stronger or Safer," *Defense One*, March 16, 2017.

[4] A part of this readiness crisis includes a personnel shortage. USAF is having trouble retaining pilots, which may in part be due to the high operations tempo but also to the availability of higher-paying jobs in the commercial airline sector. For more on the pilot shortage, see U.S. Government Accountability Office, *Military Personnel: DoD Needs to Reevaluate Fighter Pilot Workforce Requirements*, Washington, D.C.: Government Accountability Office, April 11, 2018.

[5] James N. Mattis, "Remarks by Secretary Mattis on the National Defense Strategy," U.S. Department of Defense (website), January 19, 2018.

Readiness underpins responsiveness, and both are necessary for protecting U.S. national security interests. Therefore, responsiveness—which is the time it takes for a unit to deploy to the location of a contingency and effectively carry out its operation—is a function of readiness, speed, and distance. In general, notable military scholars agree that faster is better. Sun Tzu, in particular, argued that "rapidity is the essence of war."[6] The Prussian military scholar Carl von Clausewitz agreed and argued that the second principle of war was to "act with the utmost speed. No halt or detour must be permitted without good cause."[7]

Most Americans today insist on having a very responsive military.[8] Both political leaders and the general public expect that U.S. forces can quickly and effectively carry out operations nearly anywhere in the world. Take the 2012 attack on the U.S. diplomatic compound in Benghazi for example. The failure to defend U.S. personnel created such an outcry that it led the Department of Defense (DoD) to create several rapid response teams that would be stationed near potential hotspots and prepared to prevent another such incident.[9] For its part, USAF maintains that it is a "globally responsive force—always ready" and that it counts "responsiveness in minutes and hours not weeks or months."[10] But according to the former chief of staff General Mark Welsh, the Air Force's "strategic agility and responsiveness requir[e] a high state of readiness across the Total Force."[11] Yet it is unclear how responsive of a force the United States actually needs.

Because one of DoD's single greatest expenses is its policy of maintaining a relatively high level of readiness across the force, this issue merits further exploration.[12] This report aims to inform the debate on readiness by exploring past demand for responsiveness in U.S. military operations as way of gauging what future demand might be. In doing so, it examines the following questions: When is responsiveness needed, and how is it tied to achieving political objectives? Historically, has the joint force been called on to execute multiple operations at the same time? How responsive is airpower? What factors inhibit and enable responsiveness? This study will not provide a definitive answer to these complicated questions. The intent is to identify lessons from the historical record about responsiveness that are relevant for today's

---

[6] Sun Tzu, *The Art of Warfare*, Lionel Giles, trans., New York: Race Point, 2017, p. 36.

[7] C. V. Clausewitz, *On War*, M. Howard and P. Paret, ed. and trans., Princeton, N.J.: Princeton University Press, 1976, p. 617.

[8] Historically, the opposite was true. See Betts, 1995, p. 5.

[9] New quick response forces were established in Spain, Djibouti, and Kuwait in the last several years. Jon Harper, "New Crisis Response Force Gets Ready to Deploy to Middle East," *Stars and Stripes*, September 29, 2014; Michelle Tan, "Army Quick-Response Forces Stood Up Around the World," *Air Force Times*, November 10, 2013; James K. Sanborn, "Deal with Spain Gives Marines Permanent Base, Surge Capability," *Marine Corps Times*, June 19, 2015.

[10] U.S. Air Force, *America's Air Force: A Call to the Future*, Washington, D.C., July 2014, p. 6.

[11] General Mark A. Welsh III, *Department of the Air Force Presentation to the Committee on Armed Services House of Representatives: Impact of Sequestration*, February 13, 2013.

[12] Harrison argues that "nearly every part of the defense budget is related to readiness in one form or another." But what this study focuses on is near-term readiness. Harrison, 2014, pp. 38–40.

situation. Inevitably, these insights will be tentative and will require additional, more detailed analyses to provide conclusive answers. Nonetheless, this report makes an important contribution to the existing debates, which are often based on little empirical analysis. To the extent that evidence is offered, both those calling for enhancing readiness and their skeptics tend to cherry-pick historical examples that support their position. This report is intended to be a first cut at a historical analysis that will inject additional empirical evidence into the discussion.

This historical analysis has several components. First, I examine ten cases to determine the level of responsiveness and what role, if any, responsiveness played in achieving U.S. military and political objectives. These cases, which are listed in Table 1.1, were chosen because they varied on several key dimensions, including the type and location of the operation, the date on which they occurred, and the level of responsiveness.[13] The cases, therefore, cover a fairly broad range of past demands and present differing outcomes. While the cases include a major war (Korea) and several deterrence operations, none of the cases is a deterrence operation against a great power adversary. This omission was due to the fact that at the time that study was conducted, a great power war did not seem particularly likely; nor had it been the focus of recent U.S. operations. Since that time, great power competition has reemerged as a major driver of U.S. defense policy. The relevance of the lessons from these cases for an era of great power competition will be discussed in the conclusion.

All of the cases examined were important U.S. military operations. An additional factor that affected case selection was the availability of open-source information.[14] In these cases, I identified how quickly the joint force deployed and what factors inhibited or enabled responsiveness.[15] Because this work was done for USAF, particular attention is paid to the role of airpower in each of the cases. Yet the cases tell the story of how quickly the joint force deployed. The responsiveness of different types of forces could not be examined in isolation from the entire operation because at least in part the forces deployed were affected by the pool of available forces across domains. So while airpower is at times called out in the narrative, the report is examining the demand for responsiveness within the joint force—not just the USAF.

Second, I created a database of U.S. military operations from 1945 through 2013 and identified how often the joint force has been called on to respond to multiple contingencies simultaneously. This was done because it was clear that looking at the size of the force and urgency associated with individual cases understated the potential total responsiveness required because operations can take place at the same time. Because of time and resource limitations,

---

[13] This excludes other capabilities that USAF provides, including space and cyber, which did not lend themselves to an open-source analysis.

[14] In other words, I omitted the most recent operations, such as Operation Enduring Freedom or Operation Iraqi Freedom, because little open-source data on force deployments were available.

[15] Yet the open-source historical record on many of these operations is not nearly as complete as one would hope. Timelines detailing the day-by-day deployment of every unit or ship or aircraft were often not available for more recent U.S. operations. The cases are as detailed as the historical record allows them to be.

Table 1.1. Cases Examined

Operation	Dates Examined	Location	Type of Operation
Korean War	June 30, 1950–1951	Korean peninsula	Defense
Blue Bat	July 14–October 1958	Lebanon	Foreign internal defense
Second Taiwan Strait Crisis	August 23–October 1958	Taiwan	Deterrence/defense
Nickel Grass	October 14–November 14, 1973	Israel	Military assistance (airlift)
Mayaguez	May 12–14, 1975	Cambodia/Sea of Thailand	Hostage rescue
Desert Shield	August 1990–November 1990	Saudi Arabia	Deterrence/defense
Vigilant Warrior	October 8–November 1994	Kuwait	Deterrence/defense
Deliberate Force	August 30–September 21, 1995	Bosnia	Compellence
Desert Fox	December 16–19, 1998	Iraq	Compellence
Allied Force	March 24–June 10, 1999	Kosovo	Compellence

this is a simple frequency analysis that does not include many other potentially interesting issues, such as the size of the force required or the length of the operation.

This approach is not without its limitations. The past demand for responsiveness does not necessarily reflect what is required of the USAF today or in the future. In fact, there is no way to measure the true demand for forces. This is because both past operations and current plans are imperfect measures that also reflect the available supply. In other words, military planners and commanders do not ask for more forces than exist, even if they would be useful.[16] At the same time, commanders may have an incentive to ask for the rapid deployment of more forces than may actually be needed to carry out an operation. The logic here is simple: if forces are available, a commander is likely to ask for them to reduce the risk of failure. For these reasons, there is no true or objective requirement for responsiveness. Nevertheless, surveying ten operations that have occurred since World War II does help to provide insight as to whether U.S. forces have often been called on to rapidly respond to crises or whether this is a relatively infrequent occurrence. Equally importantly, such a survey can help to identify if there are certain types of operations where responsiveness is important and roughly how large of a force has been needed.

The rest of this report proceeds as follows. The second chapter overviews the level of joint responsiveness in each of the cases. The third chapter explores the issue of simultaneity events— when the U.S. military has been called on to carry out two or more operations at the same time. The fourth chapter considers the relationship between responsiveness and the size of the force employed. The fifth chapter explores when responsiveness is necessary. The sixth chapter identifies the factors that enable or hinder a rapid military response and considers the risks of focusing on speed. The final chapter includes findings and recommendations.

---

[16] I thank Jeff Hagen for making this point.

2. Responsiveness in Ten Cases

Responsiveness can be measured in minutes, hours, days, weeks, months, or even years. For the purposes of this study, I focused on a rough measure of the time it took for the bulk of the force employed to respond (hours, days, weeks, or months). In the past, it often took the United States years to spin up the forces that it needed to prevail in a conflict; however, since World War II (the period examined in this study) its forces have maintained a much higher level of readiness.[1] As a reflection of this fact, the lowest level of responsiveness considered in this study is months. The other key component of this definition is the majority of the force deployed, which I define as when 50 percent or more of the forces used in the operation reached the theater of operations. This threshold was chosen because there was variation among the cases, and one could reasonably argue that at this threshold there were enough forces in place to carry out the operation. I considered other metrics, such as when the initial forces arrived or began to carry out operations, but found that these did not offer much insight into responsiveness. The first responders were on the scene and operating within twenty-four hours of the deployment order being issued in nearly all of the cases. Additionally, there were only a few instances where this initial wave of forces was capable of carrying out the operation on its own.

Figure 2.1 shows the level of responsiveness for each of the cases. In two of the cases—Deliberate Force and Desert Fox—the majority of the forces were in place and ready to operate within hours.[2] In the former, North Atlantic Treaty Organization (NATO) forces began to strike Bosnian Serb targets the same day that Lieutenant General Rupert Smith, the UN Protection Force Commander in Bosnia-Herzegovina, authorized the operation.[3] This was due to the fact that most of the NATO forces were already in the theater enforcing a no-fly zone over Bosnia-Herzegovina.[4] In the latter, U.S. commanders consciously decided to only use the forces already in the theater to achieve tactical surprise.[5] This was done to ensure that Saddam Hussein stopped

---

[1] Betts, 1995, pp. 19–20.

[2] Additionally, because Desert Fox was executed at the time and place of the United States' choosing, there was at least a week of warning time that facilitated the rapid response in this case. Ross Roberts, "Desert Fox: The Third Night," *Proceedings*, Vol. 125/4/1, April 1999.

[3] Christopher M. Campbell, "The Deliberate Force Air Campaign Plan," in Owen, 2000, p. 110.

[4] There had been a buildup of forces after a USAF F-16C had been shot down by a Serbian SA-6 surface-to-air missile on June 2, 1995. The number of U.S. aircraft at the beginning of Deliberate Force was 122, and it was 141 at the close of the operation on September 14, 1995. Richard L. Sargent, "Aircraft Used in Deliberate Force," in Owen, 2000, pp. 202–204.

[5] Michael Knights, *Cradle of Conflict: Iraq and the Birth of the Modern U.S. Military*, Annapolis, Md.: Naval Institute, 2005, pp. 199–200. The United States did deploy a crisis response force, but most of these forces did not arrive until after the operation had ended. Anthony H. Cordesman, *Desert Fox: Key Official U.S. and British Statements and Press Conferences*, Washington, D.C.: Center for Strategic and International Studies, January 31, 1998, p. 153.

Figure 2.1. Level of Responsiveness in Ten Cases

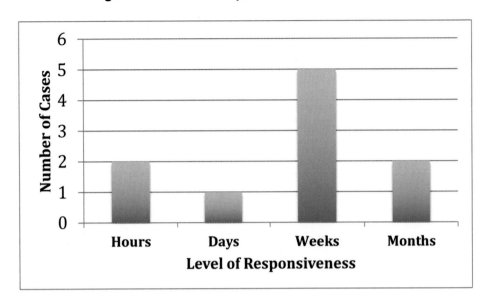

playing his game of cheat and retreat with the United Nations (UN) inspectors in which he eroded international support for the sanctions regime against Iraq by defying the UN, only to then back down on the brink of military confrontation. For instance, in November 1998, U.S. B-52 bombers en route to Iraq had been called off when Saddam Hussein complied with U.S. demands at the last possible minute. To avoid another such incident, Desert Fox relied on in-theater assets and began with an attack from the sea with a salvo of approximately 200 Tomahawk cruise missiles and carrier-based air strikes. After the first night, B-52 bombers armed with cruise missiles flying from Diego Garcia and aircraft based in Bahrain and Kuwait joined the combat operations.[6] Reinforcements had been dispatched after Desert Fox started, but few forces arrived before the four-day operation ended.[7] One exception was the USS *Carl Vinson* Battle Group, which was already en route from the western Pacific and reached the Persian Gulf in time to join the final day of strikes.[8]

There was only one case (the *Mayaguez* rescue) where the bulk of forces were ready to operate within days of the order being given. In this instance, U.S. forces were fairly responsive because they were stationed close to the Gulf of Thailand. Key forces employed included forward-based intelligence surveillance and reconnaissance aircraft in the Philippines and Thailand, Thai-based combat aircraft for air support, and around 1,000 Marines stationed in

---

[6] These forces had been deployed to the Persian Gulf on November 11 during an earlier crisis. Lt Col Paul K. White, *Crises After the Storm: An Appraisal of U.S. Air Operations in Iraq Since the Persian Gulf War*, Washington, D.C.: Washington Institute for Near East Policy, 1999, p. 57.

[7] The crisis force deployed was composed of the *Carl Vinson* battle group in addition to an air expeditionary wing consisting of around 36 combat aircraft, including F-117s. Cordesman, 1998, p. 153.

[8] White, 1999, pp. 59–60.

Okinawa and the Philippines to rescue the hostages.[9] Nevertheless, it was nearly 62 hours before U.S. ground forces were in place and ready to assault the Cambodian island of Koh Tang, where it was believed the U.S. crew was being held in addition to the *Mayaguez* itself. Moreover, because key naval forces including the aircraft carrier USS *Coral Sea* were not projected to arrive early enough, the United States chose to disobey the Thai government, which had prohibited U.S. forces based in Thailand from participating in the rescue operation. In short, U.S. forces were quite responsive, but this came at some cost politically.

The largest number of cases are ones in which responsiveness was measured in weeks, including Blue Bat, the Second Taiwan Straits Crisis, Nickel Grass, Vigilant Warrior, and Allied Force. Although U.S. military responses in all of these cases took weeks, there are some notable differences in the level of responsiveness. Because two of these cases (Blue Bat and Vigilant Warrior) included the deployment of a sizable ground force, the level of responsiveness was quite high. For instance, during the 1958 intervention in Lebanon, the deployment of over 15,000 soldiers and Marines took 25 days from the time the order had been given, but this belies the rapidity of the initial response. In less than 6 days, a composite air strike force (CASF) had deployed to Turkey, a four-battalion amphibious task force had landed in Lebanon, an Army task force had been airlifted into Lebanon, and 51 U.S. Navy (USN) ships were in the eastern Mediterranean.[10] By early August, the U.S. presence in the region peaked with 87 ships, 55 USAF aircraft, and 15,000 soldiers and Marines.[11]

Similarly, in Vigilant Warrior (1994), the first U.S. forces were on station within 24 hours, and a significant force was assembled in around 3 weeks. An amphibious ready group (ARG) that was in the United Arab Emirates (UAE) halted its exercises and moved off the coast of Kuwait within 24 hours, making it the first reinforcements to arrive. Within 2 days of the deployment order, the United States had bolstered its presence in the Persian Gulf by moving the aircraft carrier the USS *George Washington* from the Mediterranean into the Red Sea, putting 60 combat aircraft in range of striking Iraq (when supported by USAF aerial refueling). At the same time, the 1st Brigade minus the 24th Infantry Division (Mechanized) had arrived in Kuwait and was falling in on its prepositioned equipment. In another 2 days, this brigade was combat ready. USAF forces did not arrive until 72 hours after the deployment order. This delay was due to the fact that many of the forces assigned to the 23rd Composite Wing were engaged in a

---

[9] For the USAF forces in theater, see Clayton K. S. Chun, *The Last Boarding Party: The USMC and the SS* Mayaguez *1975*, Oxford: Osprey, 2011, p. 20. See also Daniel L. Haulman, "Crisis in Southeast Asia: *Mayaguez* Rescue," in Timothy Warnock, ed., *Short of War: Major USAF Contingency Operations, 1947–1997*, Washington, D.C.: Air Force History and Museums Program, 2000a, pp. 105–114; and Ralph Wetterhahn, *The Last Battle: The* Mayaguez *Incident and the End of the Vietnam War*, New York: Plume Group, 2002.

[10] George S. Dragnich, *The Lebanon Operation of 1958: A Study of the Crisis Role of the Sixth Fleet*, Arlington, Va.: Center for Naval Analysis, September 1970, pp. 77, A-1-A-8, E-2, E-3; Robert D. Little and Wilhelmine Burch, *Air Operations in the Lebanon Crisis of 1958*, USAF Historical Division Liaison Office, Washington, D.C., October 1962, declassified on February 23, 1982, p. 25.

[11] Dragnich, 1970, p. D-1; Little and Burch, 1982, pp. 25, 30–37.

Red Flag exercise in Nevada. Consequently, they had to make a double move in a very short period, first returning to their home stations and then deploying nearly 6,000 nautical miles to the Persian Gulf.[12] In 12 days, the first Army afloat prepositioned equipment arrived in the region to meet up with the soldiers who were being airlifted from the continental United States (CONUS). Within 25 days, USAF and the Civil Reserve Air Fleet had airlifted nearly 21,000 military personnel to the region.[13] In nearly every respect, Vigilant Warrior well outstripped the speed of Operation Desert Shield.

The 1973 resupply of Israel, which was primarily an air operation, was also carried out with great alacrity. Within the first five days, Military Airlift Command (MAC) had completed 100 missions, and over the span of a month USAF transport aircraft flew 567 sorties (see Figure 2.2), delivering 22,328 tons of supplies to Israel.[14] Some of this equipment was reportedly delivered to the Israeli front lines within three hours of being off-loaded at Lod Airport.[15] This accomplishment is even more striking as the airlift was constrained by the fact that there was only one en route base available to U.S. aircraft, Lajes Airbase in the Azores.

Figure 2.2. Number of Sorties Completed in Operation Nickel Grass (1973)

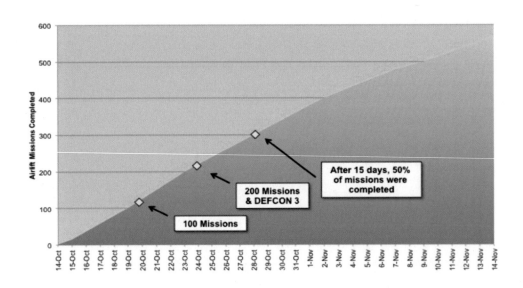

SOURCE: Kenneth L. Patchin, *Flight to Israel*, Scott AFB, Ill.: Military Airlift Command, April 30, 1974.

---

[12] DoD, Office of the Assistant Secretary of Defense (Public Affairs), "Background Briefing: Subject Iraq," October 20, 1994.

[13] U.S. Congressional Budget Office, *Moving U.S. Forces: Options for Strategic Mobility*, February 1997, pp. 4, 35.

[14] Patchin, 1974, p. 178.

[15] Patchin, 1974, p. 14.

By contrast, in the Second Taiwan Strait Crisis (1958), which was primarily a naval and air operation, the U.S. military's response lagged a bit. Given the number of forces that were based in Japan at the time, one would have expected a very quick U.S. military reaction. Moreover, because the United States had initiated a number of flexible deterrent operations in early August, there were already a handful of USAF aircraft in Taiwan, and two USN ships were continuously patrolling the Taiwan Strait.[16] After the People's Liberation Army began a large artillery bombardment of the Kinmen Islands on August 23, these aircraft carried out an air defense exercise over Taiwan.

On August 28, two aircraft carrier groups from the 7th Fleet joined the ships already patrolling the strait. Moreover, two additional aircraft carriers and their destroyer escorts from outside of the western Pacific theater were quickly dispatched. The USS *Essex*, which was in the Mediterranean, and the USS *Midway*, which was in port in Hawaii, were ordered to sail to Taiwan in the first days of the crisis.[17] By contrast, U.S. officials did not authorize the deployment of additional air forces based in Japan or CONUS for 5 days.[18] Unlike Blue Bat, where the CASF closed in Turkey in 6 days, the CASF for Taiwan took 15 days to arrive at its bases in the Pacific. This delay was due to the fact that the CONUS-based combat aircraft had to wait for tankers to get in place before they could deploy, and then they had a long journey with multiple stops that was complicated by bad weather.[19] Marine Air Group 11, which was based at Naval Air Station Atsugi in Japan, did not deploy to Taiwan for nearly 2 weeks because it encountered problems locating suitable airfields.[20] Moreover, given their starting locations (Hawaii and the Mediterranean), the aircraft carrier groups sent to reinforce the 7th Fleet took 10 days and 22 days, respectively, to arrive in the theater.[21] The last force to arrive was a Nike-Hercules battalion based in Fort Bliss, Texas. Although it was ordered to deploy as soon as the crisis erupted, the Nike battalion did not close on Taiwan until October 9, and it was not operational until the twenty-fifth of October, 2 months after the operation began.[22]

In contrast to the Taiwan crisis, there was not a great deal of urgency in Operation Allied Force (OAF) (1999). OAF was intended to be a limited coercive aerial campaign that ended up lasting much longer than expected as Yugoslav President Slobodan Milosevic only capitulated after 78 days of bombing. Because U.S. officials believed that Milosevic would agree to return to the bargaining table after a few days of airstrikes, they assumed that the modest U.S. force

---

[16] M. H. Halperin, *The 1958 Taiwan Straits Crisis: A Document History*, Santa Monica, Calif.: RAND Corporation, RM-4900-ISA, 1966, p. 65.

[17] Jacob Van Staaveren, *Air Operations in the Taiwan Crisis of 1958*, Washington, D.C.: USAF Historical Division Liaison Office, November 1962, p. 23.

[18] Van Staaveren, 1962, p. 19.

[19] Van Staaveren, 1962, p. 21.

[20] Van Staaveren, 1962, p. 24.

[21] Van Staaveren, 1962, p. 23.

[22] Van Staaveren, 1962, p. 26.

(214 aircraft) already in place was sufficient.[23] The decision to allow the USS *Theodore Roosevelt* carrier battle group (CVBG), which had been in the Mediterranean, to continue on to the Persian Gulf right before OAF began was a sign of this mindset.[24] The short war assumption, however, proved to be wrong. As the conflict dragged on, the combatant commander continued to request additional reinforcements, which were only gradually approved by the secretary of defense.[25] On April 3 the USS *Theodore Roosevelt* CVBG reversed course and returned to the Mediterranean, but it would not be on station for several more days, nearly 2 weeks into the conflict.[26] By April 10 the number of U.S. aircraft committed to OAF had nearly doubled.[27] Incrementally, additional U.S. reinforcements continued to deploy for the air war over Kosovo. Although these forces were dispatched in drips and drabs, by the time the operation ended in early June, there were 720 U.S. aircraft in place.[28] In addition to fixed wing aircraft, the deployment of an Army unit from Europe (Task Force Hawk) that included attack helicopters to Albania was authorized on April 3, when the president finally relented to the combatant commander's request. Task Force Hawk's deployment encountered numerous problems, and it ended up taking 23 days for the 55 helicopters to self-deploy from Germany to Albania.[29] Moreover, the unit was not fully operational until May 7.[30]

The operations that had the lowest level of responsiveness (months) were also the largest: the Korean War and Operation Desert Shield. It should come as little surprise that U.S. forces were least responsive in the Korean War. Not only were the understrength U.S. units initially deployed largely ineffective, but deployment timelines were quite lengthy even for forces in the theater.

---

[23] Ivo H. Daalder and Michael E. O'Hanlon, *Winning Ugly: NATO's War to Save Kosovo*, Washington, D.C.: Brookings Institution, 2000, p. 2; DoD, *Report to Congress, Kosovo/Operation Allied Force After Action Report*, Washington, D.C., January 31, 2000, p. 31. The number of aircraft did not include the B-2s, which launched strike operations from the United States. There were also 130 allied aircraft in theater.

[24] Bruce R. Nardulli et al., *Disjointed War: Military Operations in Kosovo, 1999*, Santa Monica, Calif.: RAND, MR-1406-A, 2002, p. 31.

[25] For instance, on April 13, Clark asked for an additional 300 U.S. aircraft, but these were only approved incrementally, and not all of them had been deployed when the operation finally ended. Nardulli et al., 2002, pp. 32–33. There is a separate question about whether U.S. European Command needed as many forces as Clark asked for. Sortie rates were often low due to bad weather, not because of limited force structure. Benjamin S. Lambeth, *NATO's Air War for Kosovo: A Strategic and Operational Assessment*, Santa Monica, Calif.: RAND, MR-1365-AF, 2001, p. 37.

[26] Lambeth, 2001, p. 30. The Clinton administration could have also ordered the USS *Enterprise*, which was returning from the Persian Gulf at the time, but chose not to extend its deployment. H. H. Gaffney, *Warning Time for U.S. Forces' Responses to Situations*, Alexandria, Va.: Center for Naval Analysis, June 2002, p. 8.

[27] On April 10, DoD reported that there were about 400 U.S. aircraft participating in OAF. By April 16, that number had grown to 463. DoD, Office of the Assistant Secretary of Defense (Public Affairs), "DoD News Briefing, Saturday April 10, 1999 2:05 pm," April 10, 1999; DoD, Office of the Assistant Secretary of Defense (Public Affairs), "DoD News Briefing, Friday April 16, 1999," April 16, 1999.

[28] DoD, Office of the Assistant Secretary of Defense (Public Affairs), "DoD News Briefing, Wednesday June 2, 1999 2:10 pm," June 2, 1999.

[29] Nardulli et al., 2002, pp. 70–72.

[30] Nardulli et al., 2002, p. 75. Task Force Hawk was never employed in combat.

Because the United States was in the midst of a massive post–World War II demobilization, it had great difficulty quickly mounting a rapid and effective defense of South Korea. The most that its forces could do was to hold a line around the city of Pusan, preventing South Korea from collapsing entirely. In the first month of the war, few U.S. reinforcements arrived from outside of the East Asian theater beyond the two bomber groups that were deployed from CONUS to Japan.[31] Most of the Pacific fleet was stationed on the West Coast of the United States, which meant that additional ships were slow to arrive.[32] This was also true of CONUS-based ground forces. For instance, the first elements of the 2nd Infantry Division sailed for Japan on July 17, but they did not arrive in theater until July 31, and their deployment was not complete until August 20.[33]

Even the intratheater movement of forces, primarily from Japan to Korea, proceeded slowly. The entire 24th Division had been ordered to deploy to Korea, but it was dispersed across the Japanese islands, and there were no ships immediately available to transport the ground troops.[34] The deployment, therefore, proceeded incrementally, and the first ground unit to arrive in Korea was the ill-fated Task Force Smith. Task Force Smith was ordered to delay the advancing North Korean forces to buy time for additional American reinforcements to arrive, but the undersized infantry unit, consisting of only 406 soldiers, was quickly routed by the superior enemy force, which included tanks.[35] The size and composition of Task Force Smith had been determined by the number of C-54 transport aircraft that were immediately available in Japan.[36] Because the rest of the 24th Division was transported by sea, the unit did not close in Korea until July 5, a month after Task Force Smith had been defeated in battle.[37] In short, it took the United States months to assemble a force capable of going on the offensive against the North Koreans.

By contrast, when Saddam Hussein invaded Kuwait in August 1990, initially U.S. forces were not tasked with rolling back the Iraqi forces. Instead Operation Desert Shield was intended

---

[31] Robert Frank Futrell, *The United States Air Force in Korea, 1950–1953,* Washington, D.C.: Office of Air Force History, 1983, pp. 73–74.

[32] James A. Field Jr., *History of United States Naval Operations: Korea,* Washington, D.C.: U.S. Department of the Navy, 1962. The first aircraft carrier, the USS *Boxer,* did not dock at Yokosuka until July 23. A. Timothy Warnock, ed., *The U.S. Air Force's First War: Korea 1950–1953 Significant Events,* Montgomery, Ala.: Air Force History and Museums Program, Air Force Historical Research Agency, 2000, p. 5.

[33] Terrence J. Gough, *U.S. Army Mobilization and Logistics in the Korea War: A Research Approach,* Washington, D.C.: Center of Military History, U.S. Army, 1987, pp. 4–6.

[34] T. R. Fehrenbach, *This Kind of War,* Washington, D.C.: Brassey's, 1963, p. 65.

[35] Task Force Smith was ordered to confront and delay the advancing North Korean forces to buy time for additional American reinforcements to arrive, but it was inadequate for this mission and was routed in several hours. Task Force Smith consisted of two understrength infantry companies along with some headquarters and communications personnel, totaling 406 soldiers. They were armed with their rifles and two 75-mm recoilless rifles, two 4.2-in mortars, six 2.36-in rocket launchers, and four 60-mm mortars. They were later joined by part of an artillery unit with six 105-mm howitzers. Roy E. Appleman, *South to the Naktong, North to the Yalu (June–November 1950),* Washington, D.C.: Center of Military History, U.S. Army, 1992, p. 61; Fehrenbach, 1963, pp. 65–66.

[36] John Garrett, *Task Force Smith: The Lesson Never Learned,* Fort Leavenworth, Kan.: School of Advanced Military Studies, U.S. Army Command and General Staff College, 2000, p. 6.

[37] Garrett, 2000, p. 7.

to deter the Iraqi dictator from attacking Saudi Arabia and, if deterrence failed, defending the gulf kingdom.[38] Once the order to begin Desert Shield was issued on August 7, the United States quickly moved a joint force into the theater as a deterrent to further Iraqi aggression. In the first 24 hours, sea-based air forces began to arrive in the Persian Gulf. First, the aircraft carrier the USS *Independence*, which had been sailing toward the region since August 4, arrived on station in the Gulf of Oman, while the USS *Eisenhower* CVBG crossed through the Suez Canal entering the theater. The following day, the first forces from CONUS, the 71st Tactical Fighter Squadron and several E-3s, landed at Dhahran airbase in Saudi Arabia and immediately began defensive combat air patrols. After another 24 hours passed, a second squadron of F-15C aircraft touched down in Saudi Arabia, and the first elements of the 82nd Airborne Division also landed in the kingdom and secured the perimeter around Dhahran Airport. Eight days after the deployment order was issued, the entire Second Brigade of the 82nd Airborne was in place, 245 USAF aircraft were in the theater (including 134 fighters), and advance elements of the Seventh Marine Expeditionary Brigade (MEB) had disembarked at Jubayl, Saudi Arabia.[39]

Although the United States quickly deployed light ground-based units to Saudi Arabia, this force was little more than a trip wire. As seen in Figure 2.3, ground forces did not arrive in mass (particularly armored units) for at least another 30 days.

Figure 2.3. Army Units Deployed in Desert Shield

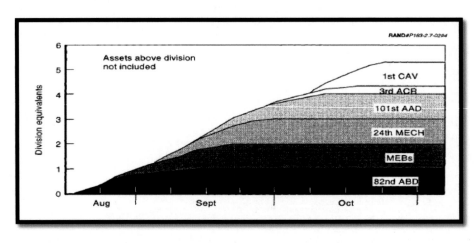

SOURCE: *Project AIR FORCE Assessment of Operation Desert Shield: The Buildup of Combat Power*, Santa Monica, Calif.: RAND Corporation, MR-356-AF, 1994, p. 13.
NOTES: 82nd ABD: 82nd Airborne Division; MEBs: Marine Expeditionary Brigades; 24th MECH: 24th Mechanized Division; 101st AAD: 101st Airborne Division; 3rd ACB: 3rd Armored Cavalry Regiment; and 1st CAV: 1st Cavalry Division.

---

[38] The forces used for Operation Desert Shield were later employed to liberate Kuwait, but that decision was not made until late October and implemented in the phase 2 deployment, which is not considered here. As the United States initiated operations against Iraqi forces at the time and place of its choosing, responsiveness was not particularly important.

[39] Emery M. Kiraly, Robert C. Owen, and Aron Pinker, *Part II: Chronology of the Gulf War*, in Eliot A. Cohen, ed., *Gulf War Airpower Survey*, Vol. 5, Washington, D.C.: U.S. Department of Defense, 1993, pp. 9–21.

Similarly, USN units that had to deploy from CONUS or other distant theaters were not on station for several weeks, if not longer.[40] By contrast, USAF aircraft continued to rapidly arrive in the Persian Gulf (see Figure 2.4). Within 16 days, 384 aircraft, which was half of the force allocated for the initial deployment (through November), had arrived in the Persian Gulf.

Figure 2.4. U.S. Air Force Deployments, Operation Desert Shield, August 7–November 9, 1990

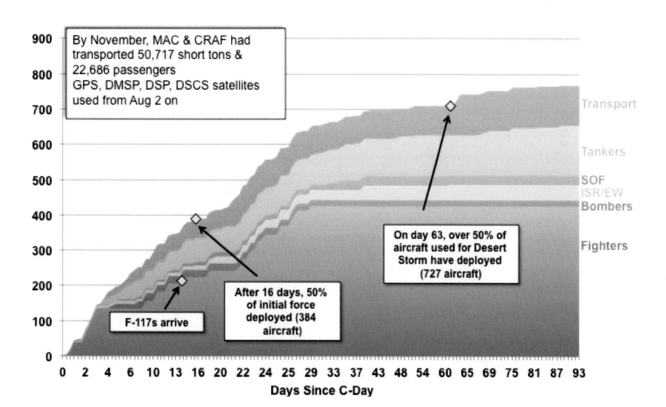

SOURCE: Eliot A. Cohen, ed., *Gulf War Airpower Survey*, Vol. 5, Washington, D.C.: U.S. Department of Defense, 1993.
NOTES: MAC: military airlift command; CRAF: Civil Reserve Air Fleet; GPS: Global Positioning System; DMSP: defense meteorological satellite program; DSP: Defense Support Program: DSCS Defense Satellite Communications System; C-day: the day a deployment operation begins; SOF: Special Operations Forces; and ISR/EW: Intelligence Surveillance and Reconnaissance/Electronic Warfare.

---

[40] For instance, other major naval combatants, including the *Saratoga* and *Wisconsin*, arrived in late August, while the ARG carrying 2/4 MEB and *Kennedy* not until mid-September. DoD, *Conduct of the Persian Gulf War: Final Report to Congress*, Washington, D.C.: Office of the Secretary of Defense, April 1992, p. E-24.

3. The Demand for Simultaneous Operations

During the case study analysis, it became apparent that looking at the demand for responsiveness in one operation at a time potentially understates the true demand for responsiveness because U.S. forces may be called on to respond to multiple contingencies simultaneously. I call this a simultaneity event—that is, a situation in which the U.S. military was engaged in two or more operations at the same time. For instance, in 1958 the U.S. military was ordered to undertake interventions in the Middle East and East Asia at the same time. The Middle Eastern crisis, which precipitated Operation Blue Bat, began on July 14, when the pro-Western Iraqi monarchy was overthrown in a coup that brought a pro-Nasserist government to power. In approximately 3 weeks, the United States deployed 55 aircraft, 87 ships, and 15,000 ground forces in response to the Lebanese government's request for assistance. This operation was still ongoing when on August 23, the People's Republic of China (PRC) began a large bombardment of Taiwan's offshore islands. Some U.S. policymakers saw these two events as linked and feared that the communist block was trying to exploit the fact that the United States was focused on the Middle East.[1] The Eisenhower administration decided that it must assemble a large force to deter China from seizing the islands and to strengthen deterrence globally.[2] Consequently, the United States deployed 150 ground-based aircraft and 4 additional aircraft carriers in 26 days to East Asia.

This example demonstrates that simultaneous operations have occurred and that being highly responsive to multiple crisis can be important. In seeking to justify its readiness model, USAF has argued that it must be prepared to rapidly conduct simultaneous operations and that reduced readiness would imperil its ability to do so. A more recent example it has pointed to is March Madness 2011.[3] At that time, the emergence of two simultaneous crises in addition to ongoing U.S. military operations stretched USAF's capacity to its limits. During this period, USAF continued to provide support to the wars in Iraq (which was winding down) and Afghanistan (which was in the midst of a surge), counterterrorism operations in Africa and the Middle East, and counternarcotics operations in Latin America. Unexpectedly, USAF also led an air campaign against Libya and participated in extensive disaster relief efforts in Japan. A critical question is whether March Madness is the exception or the rule.

As a first step toward answering this question, I explored the frequency of simultaneity events. To identify simultaneity events, I created a list of significant U.S. military operations

---

[1] Van Staaveren, 1962, p. 14.

[2] Halperin, 1966, pp. 121–122.

[3] General Norton Schwartz, "Air Force Contributions to Our Military and Our Nation," speech delivered at the World Affairs Council of Delaware, Wilmington, Del., January 20, 2012.

from 1947 to 2000. The foundation of this list was a previously built contingency access database.[4] Since the access database only included instances in which the United States had requested permission to use another country's territory or airspace, it needed to be augmented to ensure that it captured all significant operations, especially USN and U.S. Marine Corps (USMC) operations that might have been excluded because they did not require access to foreign territory.[5] Then, extremely small and unimportant operations (e.g., the ferrying of one plane's load of humanitarian assistance) were eliminated from the list. This made possible a conservative estimate of the number of simultaneity operations and ensured that the data were not biased by insignificant often one-off missions.

These data were then inputted into a timeline software so that it was possible to identify when two or more operations overlapped at a point in time. Each unique combination of two or more concurrent operations was counted as a simultaneity event. For instance, if two operations were being executed simultaneously and a third began but then ended while the other two continued, this was counted as two separate simultaneity events.

This analysis revealed that simultaneity events have been quite common. Between 1945 and 2000, there were 375 simultaneity events, shown in Figure 3.1. Of these simultaneity events, 247 took place during the Cold War, with an average of nearly 6 separate simultaneity events occurring each year. In particular, the early 1960s and 1980s were periods when multiple simultaneous operations occurred frequently—a phenomenon that increased further during the 1990s.

It is worth noting, however, that the simultaneity events during the Cold War tended to be composed of relatively few operations (an average of between 2 and 5), as shown in Figure 3.2.[6] By contrast, in the 1990s, simultaneity events not only occurred more frequently (an average of 11 per year), but each individual simultaneity event generally consisted of a larger number of total operations (an average of between 6 and 9). Although the frequency of simultaneity events

---

[4] Stacie L. Pettyjohn and Jennifer Kavanagh, *Access Granted: Political Challenges to the U.S. Overseas Military Presence, 1945–2014*, Santa Monica, Calif.: RAND, RR-1339-AF, 2016.

[5] Robert Mahoney, *U.S. Navy Responses to International Incidents and Crises, 1955–1975: Survey of Navy Crisis Operations*, Alexandria, Va.: Center for Naval Analysis, 1977; Adam B. Siegel, *The Use of Naval Forces in the Post-War Era: U.S. Navy and U.S. Marine Corps Crisis Response Activity, 1946–1990*, Alexandria, Va.: Center for Naval Analysis, 1991; Adam B. Siegel, *A Chronology of U.S. Marine Corps Humanitarian Assistance and Peace Operations*, Alexandria, Va.: Center for Naval Analysis, September 1994; W. Eugene Cobble, H. H. Gaffney, and Dmitry Gorenburg, *For the Record: All U.S. Forces' Responses to Situations, 1970–2000 (with Additions Covering 2000-2003)*, Alexandria, Va.: Center for Naval Analysis, June 2003; Larissa Forster, *Influence Without Boots on the Ground: Seaborne Crisis Response*, Newport, R.I.: Naval War College Press, 2010; Thomas Barnett and Linda Lancaster, *Answering the 9-1-1 Call: U.S. Military and Naval Crisis Response Activity, 1977–1991*, Alexandria, Va.: Center for Naval Analysis, 1992; Michael Butler, "U.S. Military Intervention in Crisis, 1945–1994," *Journal of Conflict Resolution*, Vol. 47, No. 2, 2003, pp. 226–248; Hank Gaffney et al., *Employment of Amphibious MEUs in National Responses to Situations*, Alexandria, Va.: Center for Naval Analysis, 2006.

[6] The scatterplot depicts a point for each year between 1947 and 2000 when there was at least one simultaneity event. The position on the horizontal access represents the total number of simultaneity events for a particular year, while the vertical access depicts the average size of those simultaneity events.

Figure 3.1. Simultaneity Events by Year

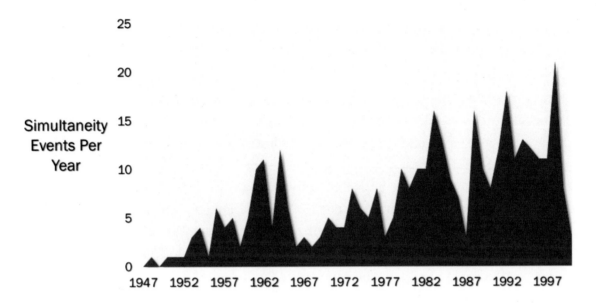

Figure 3.2. Number and Size of Simultaneity Events by Decade

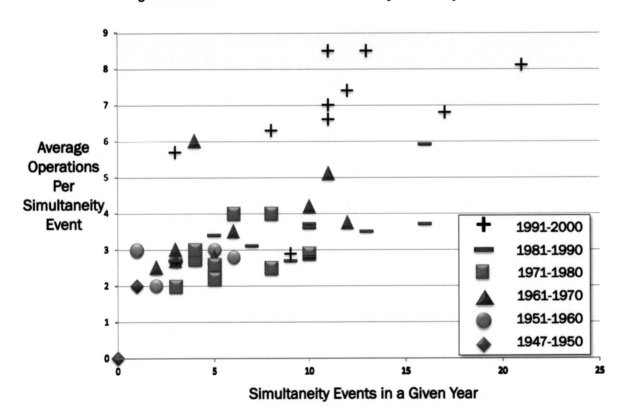

increased during the 1990s, this does not necessarily mean that the overall demands on force structure and readiness were greater.

This initial analysis cannot specify the level of readiness necessary to address concurrent military operations either within USAF or across DoD; nor can a historical analysis accurately predict the future. Nevertheless, it does suggest that simultaneous operations have been a common and increasingly frequent phenomenon that must be taken into account in future readiness needs assessments and that USAF does require a relatively high level of readiness to be able to deal with these situations. In short, given how common simultaneous operations have been in the past, USAF should expect to be called on to carry out multiple operations concurrently.

4. Responsiveness and Force Size

Not surprisingly, the level of responsiveness is closely tied to the size of the force that is deployed. It is much easier to have a small force ready and to rapidly deploy it than a large one. The notional chart in Figure 4.1 helps to illustrate this point. It shows the approximate level of responsiveness along the horizontal axis and the approximate size of the joint force employed on the vertical axis. There is clearly a direct relationship between these two factors: the more troops that are needed, the longer it takes to have them in place. This is particularly true of heavy ground units. One can imagine that in the bottom left-hand corner you have a case where extremely high responsiveness is required. An example of this would be a Predator (MQ-1) or Reaper (MQ-9) strike against a moving terrorist. In situations where the United States has sound intelligence on a mobile target, action is required in minutes, not hours, but the size of the force is quite small. Conversely, when the United States is preparing to fight a major war that includes a large number of ground forces, such as in Korea or Operation Desert Storm, it has tended to take months if not years to prepare, assemble, and deploy such a large force.

Figure 4.1. Approximate Level of Responsiveness and Size of Force Employed

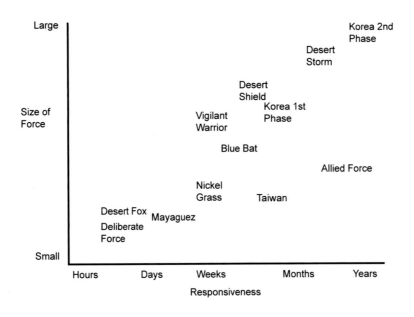

At the high end, there are deterrent and defensive operations, such as Desert Shield and the first phase of the Korean War. In these situations, there was initially great urgency, but the exact size of the force required varied depending on the circumstances. One can also question whether all of the forces deployed were necessary to achieve the United States' objectives. For instance, the United States deployed approximately 1,000 aircraft, 60 ships, an amphibious task force, and

240,000 personnel in the first phase of the Desert Shield deployment, but it is unclear whether all of these forces were needed to dissuade Saddam Hussein from turning on Saudi Arabia.[1] It is also unclear whether the United States needed to deploy over 20,000 troops for Vigilant Warrior. On October 10, 1994, when the first U.S. land-based aircraft and the lead elements of the 24th Infantry Division were arriving, Iraq announced that its forces that had assembled near Kuwait were moving northward, away from the border.[2] It appears, therefore, that the announcement the United States planned to deploy a large force to the region and the arrival of first-wave reinforcements were sufficient to convince Saddam that a second invasion of Kuwait was not likely to succeed.

Figure 4.2 shows the deployment timelines for aircraft in the cases.[3] The horizontal axis shows the number of days after the deployment order, while the vertical axis shows the number

Figure 4.2. Level of Responsiveness and the Number of Aircraft Employed

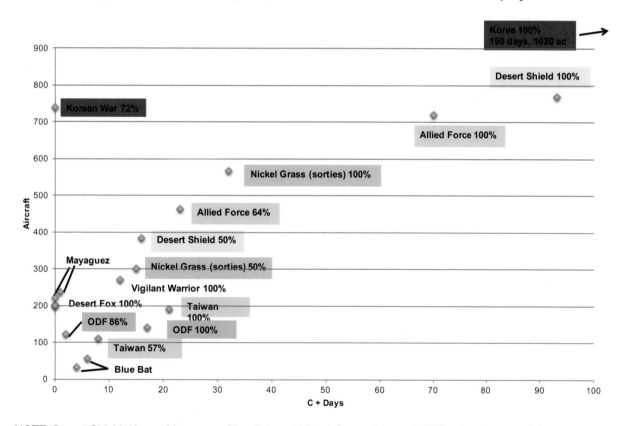

NOTE: Desert Shield, Korea, Mayaguez, Blue Bat, and Nickel Grass data are USAF only; others are joint. ODF: Operation Deliberate Force; ac: aircraft.

---

[1] DoD, 1992, p. E-17.

[2] W. Eric Herr, *Operation Vigilant Warrior: Conventional Deterrence Theory, Doctrine, and Practice*, Maxwell Air Force Base, Ala.: School of Advanced Airpower Studies, June 1996, p. 30.

[3] The data are not even across the cases. Half of the cases (Desert Shield, Korea, *Mayaguez*, Blue Bat, and Nickel Grass) only include USAF aircraft, while the other five cases include the number of joint aircraft deployed.

of aircraft deployed. For each operation, I have identified when at least 50 percent of the aircraft used were in place and when the entire force was in place.[4] The chart affirms that there is a direct relationship between the number of aircraft and the time that it takes for them to deploy. Airpower was typically quite responsive; in nearly all of the cases, 50 percent of the force was in place within three weeks. The outlier cases are the Korean War and OAF.

In most of the cases, responsiveness was facilitated by the presence of forces already in the theater. At the same time, in Vigilant Warrior, the Taiwan crisis, Blue Bat, Nickel Grass, and Desert Shield, the United States rapidly deployed aircraft from CONUS. Airpower is highly responsive because of the inherent range and speed of aircraft, which allow large numbers of them to respond to a crisis almost anywhere in the world in several weeks. By contrast, having forces in position or near the location of a contingency matters much more for ground and naval forces because of their longer transit times. Large USN forces can be assembled quickly if they are in the vicinity of a crisis, as occurred in Operation Blue Bat. But if they are not well positioned to respond, as was the case in the Second Taiwan Crisis, it generally takes them several weeks to a month to get on station. Large ground forces are slower to deploy, mainly because their heavy equipment needs to be transported by sea. However, if equipment is prepositioned at the site of a potential crisis, responsiveness is dramatically heightened, as was evidenced in Operation Vigilant Warrior, because the personnel can be flown in relatively quickly. In short, heavy ground and naval forces have greater difficulty responding quickly if the forces that are needed are far from the theater of operations compared with airpower.

This chapter highlighted the relationship between responsiveness and size of the force employed, but it did not consider the equally important question of whether responsiveness was necessary for achieving the United States' objectives. The next chapter explores this issue.

---

[4] The Nickel Grass entry represents the number of airlift sorties completed, not the number of aircraft deployed.

5. When Is Responsiveness Needed?

Responsiveness was at least moderately important in nine of the ten cases examined (see Table 5.1). This assessment is based on whether timeliness was tied to achieving the political and operational objectives. The responses are characterized as immediate (hours), rapid (days), measured (weeks), and delayed (months).[1] In some instances these political and operational aims were at odds, resulting in delay. For instance, in Kosovo, one could argue that it was important to defeat Milosevic's forces quickly to stop the killing of Kosovar Albanians, which not only continued but accelerated during the campaign.[2] Yet it took considerable time to build enough support within the North Atlantic Council (NAC) for NATO to even authorize the use of military force, and this fragile consensus was predicated on the terms that targets underwent a lengthy approval process and that the minimal level of force possible to achieve the Alliance's goals would be employed.

In six of the cases, there was great urgency because the United States was reacting to belligerence and aiming to either deter further aggression or to come to the assistance of a beleaguered partner or ally. For instance, the South Koreans needed U.S. reinforcements to halt the North Korean advance. The remnants of the Republic of Korea (ROK) army and the initial U.S. reinforcements barely held onto a defensive line around the city of Pusan in August and September 1950. While the situation was not that dire in the 1973 Arab-Israeli War, the Israeli Defense Forces (IDF) were outnumbered, faced an assault on two fronts, and only had two weeks of ammunition left at one point.[3] To meet Israel's immediate requirement for supplies, the United States had little choice but to airlift them, as it would take more than a month to get the armaments to Israel by sea.

In three cases (Second Taiwan Straits Crisis, Desert Shield, and Vigilant Warrior) the United States aimed to deter further aggression. Deterrence is only possible if a country has not yet undertaken the action to be deterred, and in each of these cases there were signs that the adversary was preparing to go on the offensive. In the case of Taiwan, the Chinese had already begun limited offensive military operations against Taiwan's offshore islands, including shelling and implementing a blockade. The Eisenhower administration sought to prevent this from becoming the precursor to a hostile annexation of the islands. In Desert Shield, Hussein had invaded Kuwait, but because his forces faced essentially no resistance, there was a fear that they

---

[1] Similar to the typology used by Karl Mueller, "Flexible Power Projection for a Dynamic World: Exploiting the Potential of Air Power," in Cindy Williams, ed., *Holding the Line: US Defense Alternatives for the Early 21st Century*, Cambridge, Mass.: MIT Press, 2001, pp. 223–224.

[2] Barry R. Posen, "The War for Kosovo: Serbia's Political-Military Strategy," *International Security*, Vol. 24, No. 4, Spring 2000, p. 63.

[3] Patchin, 1974, p. 6.

would turn west and seize valuable oil fields in the Kingdom of Saudi Arabia (KSA), but the George H. W. Bush administration had not yet decided to roll back the invasion of Kuwait. The initial response was intended to ensure that Iraqi forces massing near Saudi Arabia's eastern border did not cross over and seize territory that included valuable oil fields. Over time an additional objective was added: the coalition assembled hoped that its military buildup would also compel Saddam to withdraw from Kuwait. This goal proved to be unattainable. It is not clear whether a faster buildup would have impressed Hussein with its speed and decisiveness, forcing him to back down, as it later did in 1994.[4] In Operation Vigilant Warrior, Hussein had only begun to mobilize and mass forces along the border with Kuwait but had not yet invaded its territory when the U.S. rapidly deployed reinforcements to the region. In none of these cases is it known whether the aggressor planned to launch a full-scale invasion. But in all cases the U.S. rapidly deployed air, maritime, or light ground forces within days and then proceeded to augment its military capability in the theater over a period of weeks, resulting in deterrence holding.

Time was also of the essence during the *Mayaguez* operation. U.S. forces needed to locate the hijacked vessel before its crew were taken ashore, which would have greatly increased the difficulty of rescuing them. In 1968 U.S. forces had failed to intercept the USS *Pueblo*, a USN intelligence vessel that had been attacked by North Koreans, before its crew was relocated, resulting in a nearly year-long hostage crisis.[5]

By contrast, in Operation Blue Bat, Deliberate Force, and Allied Force, there was a middling level of urgency. In Blue Bat, President Dwight D. Eisenhower ultimately determined that the United States had to respond affirmatively to Lebanese President Camille Chamoun's request for help in order to protect U.S. credibility in the Middle East and the world.[6] A rebellion had broken out against Chamoun in May but had by July—when Blue Bat began—largely petered out. U.S. policymakers saw little immediate threat to the Lebanese government.[7] Nevertheless, they believed that the United States had to act quickly to reassure its allies around the globe. While there was no pressing operational reason to quickly intervene in Lebanon, U.S. forces had been preparing for such a contingency, and therefore many were well positioned to quickly respond. Once the president was informed that a battalion of Marines could land within 24 hours of receiving Chamoun's request, he ordered the operation to be executed on that timeline to ensure

---

[4] For more on this counterfactual, see Janice Gross Stein, "Deterrence and Compellence in the Gulf, 1990–1991: A Failed or Impossible Task," *International Security*, Vol. 17, No. 2, Fall 1992, pp. 147–179.

[5] The *Pueblo* was a USN intelligence ship that was seized by DPRK in international waters on January 23, 1968. The crew was taken to Wonson, North Korea, making the prospect of a rescue operation highly risky. President Lyndon B. Johnson ordered no direct response but began a military buildup as negotiations were ongoing. For more on the *Pueblo*, see Mitchell B. Lerner, *The* Pueblo *Incident: A Spy Ship and the Failure of American Foreign Policy*, Lawrence: University Press of Kansas, 2002; and Trevor Armbrister, *A Matter of Accountability: The True Story of the Pueblo Affair*, New York: Lyons, 2004.

[6] Salim Yaqub, *Containing Arab Nationalism: The Eisenhower Doctrine and the Middle East*, Chapel Hill: University of North Carolina Press, 2004, pp. 224–226.

[7] Little and Burch, 1962, p. 13.

that the United States' credibility did not suffer. It is quite difficult to measure reputational effects, but it is unlikely that the United States' reputation would have been seriously damaged had U.S. forces waited several days or even weeks before intervening.

In Bosnia, the United States aimed to punish the Bosnian Serbs for violating the UN safe zones and to compel them to sue for peace. The event that precipitated the NATO intervention was a mortar landing in a marketplace in Sarajevo, killing 38 civilians.[8] Yet there was no immediate and visible threat to civilians or UN peacekeepers that NATO forces had to neutralize. Instead, the most pressing reason for action was to restore NATO and the UN's credibility before the consensus for action within the two international organizations collapsed. One could argue that there was a greater need for a rapid response once the ethnic cleansing began and that an expeditious and forceful response would have brought it to an end and prevented the mortar attack in Sarajevo, as well as the attacks on the UN safe havens of Srebenica and Zepa.[9]

In OAF—the air war against Serbia—NATO intervened for humanitarian purposes and thought that a short demonstration of the alliance's willingness to use force would impel Serbia to seek a political settlement to the conflict. In short, the air campaign was seen as a last-ditch bargaining tactic. NATO had issued the activation order (ACTORD) six months earlier in October but deferred beginning the phased strike operations in the hopes that a deal would be reached.[10] It was only after months of failed negotiations that NATO finally decided to use force. While NATO chose the time that the operation would begin, its stated aim was to end the violence, which it arguably failed to do, at least initially. Between March 31 and April 8, there was the largest exodus (some 400,000) of Kosovar Albanian refugees during the entire conflict.[11] The air campaign—which was hampered by bad weather, its inability to destroy Serbia's air defenses, and the Serbian forces' use of dispersal tactics—proved unable to significantly limit the Serbian ground forces' freedom of action.[12] Over time, however, the fact that NATO remained unified, that additional air forces continued to flow into the theater (more than 700 aircraft), and that the list of approved targets was expanding to include strategic targets in Serbia seems to have pushed Milosevic toward negotiations.[13]

---

[8] Karl P. Mueller, "The Demise of Yugoslavia and the Destruction of Bosnia: Strategic Causes, Effects, and Responses," in Owen, 2000, pp. 20–21.

[9] Micah Zenko, "Saving Lives with Speed: Using Rapidly Deployable Forces for Genocide Prevention," *Defense and Security Analysis*, Vol. 20, No. 1, March 2004, pp. 3–19.

[10] DoD, 2000, p. A-3.

[11] Posen, 2000, p. 63.

[12] Posen, 2000, pp. 62–66.

[13] Posen, 2000, pp. 72–73. There were other factors at play as well, including Russian support. See Posen, 2000, pp. 66–75; Lambeth, 2001, pp. 67–86.

Finally, in one of the cases (Desert Fox) there was little urgency. This was a case in which vital U.S. interests were not at stake, and instead the United States employed force in a coercive manner. In Operation Desert Fox, President William Clinton sought to punish Saddam Hussein for not complying with UN inspections. By late 1998, Saddam had been playing games with the UN inspectors for nearly a decade, and the Clinton administration had lost patience with him. Nevertheless, there was no pressing need that the United States had to respond in December other than there was a provocative action (of which there had been many) that justified the operation.

In sum, not surprisingly the cases reveal that there is typically greater urgency when the United States is responding to the actions of others rather than dictating them. In the last several decades, the United States has become accustomed to using force punitively or coercively, in circumstances where urgency has only arisen if there were atrocities being committed.[14]

Future contingencies may require more rapid and coherent responses, for example, in responding to an enemy surprise attack.[15] Reliance on warning to reduce the need for rapid response is rendered perhaps more hopeful than realistic by the evidence that policymakers frequently have ample warning, but they misread it or choose not to act.[16] This certainly was true in Korea, the Second Taiwan Strait Crisis, and Desert Shield.

While responsiveness seemed generally important in most of the cases examined in this study, the consequences of failure were not uniform. In particular, the outcomes of small-scale contingencies (SSCs) are much less significant than the outcome of a major war.[17] In SSCs, the costs of failure are, by definition, limited. These types of operations historically have been the purview of USMC, which has long served as the United States' rapid response force, protecting U.S. personnel and interests abroad. SSCs may result in the death of U.S. and other innocent civilians and the destruction of property, but they do not pose an existential threat to the nation.

By contrast, major wars involve interstate aggression, which has the potential to cause widespread destruction and to reverberate throughout the broader international community. The

---

[14] For instance, in Operation Odyssey Dawn—the 2011 U.S. intervention in Libya—the use of force was discretionary, but there was urgency because without intervention it appeared as if rebel forces in Benghazi would have been slaughtered by Qaddafi's troops. Christopher S. Chivvis, *Toppling Qaddafi: Libya and the Limits of Liberal Intervention*, New York: Cambridge University Press, 2014, pp. 53–54; Karl P. Mueller, ed., *Precision and Purpose: Airpower in the Libyan Civil War*, Santa Monica, Calif.: RAND, RR-676-AF, 2015, p.1.

[15] Vigilant Warrior came close to a bolt-from-the-blue move because U.S. intelligence collection efforts had been seriously hampered by cloud cover for over a week prior to the crisis. Gaffney, 2002, p. 5. For more on recent responses to natural disasters in Asia, see Jennifer D. P. Moroney et al., *Lessons from Department of Defense Disaster Relief Efforts in the Asia-Pacific Region*, Santa Monica, Calif.: RAND, RR-146-OSD, 2013.

[16] Richard K. Betts, "Surprise Attack: NATO's Political Vulnerability," *International Security*, Vol. 5, No. 4, Spring 1981, p. 117.

[17] SSCs are operations short of major war, including noncombatant evacuations, humanitarian assistance, disaster relief, limited strikes, counterdrug, no-fly zones, peacekeeping, limited interventions, hostage rescue, counterpiracy, and limited counterterrorist operations.

delayed response in the Korean War, for instance, cost tens of thousands of lives and raised concerns that communist aggression was imminent worldwide, prompting the U.S. military buildup in Western Europe. Typically, important national interests are also at stake in major wars, which means that losing or even appearing at risk of failure puts those interests at risk.

In situations when the U.S. could not respond quickly to deter or prevent an attack from succeeding, it has slowly built up its forces and launched a counterattack, as was done in Korea and Iraq. However, liberation ends up being incredibly costly for the occupied countries, and adversaries have learned that they cannot prevail against the U.S. military if it is allowed to mass its superior forces and then take the offensive in its own time. Consequently, nations such as China and Russia have developed long-range precision strike capabilities and air defenses that threaten to make it extremely costly and difficult for the United States to deploy to and operate in their vicinity. In other words, the costs of a counteroffensive may be growing, making responsiveness more critical for deterrence to hold.

While the historical analysis included cases of major regional war, it did not include instances of deterring or fighting a great power war. Responsiveness is also critically important in this situation. During the Cold War, for example, DoD invested considerable time and effort so that it could quickly bolster NATO's defenses in Western Europe. CONUS-based U.S. reinforcements were kept ready and practiced rapidly deploying to Europe in Return of Forces to Europe exercises. It is important to note that the U.S. military cannot be expected to rapidly respond to all crises across the globe. U.S. policymakers have to decide whether they prioritize preserving U.S. combat capability and maintaining a high level of unit readiness so that U.S. forces can respond to a large-scale crisis or acting as the global police force.

Table 5.1 lists the ten cases examined, the purpose of each operation, the type of operation, response, outcome, risks of rapid employment, enablers of responsiveness, and hindrances. The urgency and how it is tied to mission success was discussed in this chapter. The next chapter explores the factors that enable or impede responsiveness and the risks associated with rapid deployment.

Table 5.1. Summary of Cases

Case	Purpose	Type of Operation	Response	Outcome	Risks of Rapid Employment	Enablers of Responsiveness	Hindrances
Korean War 1950	Prevent imminent defeat of ROK	Regional war	Early response ineffective/ delayed	Near loss; but after three months, counterattack drove Democratic People's Republic of Korea (DPRK) back	Forces in the region not best suited for operating environment; unready units	Forward-based forces; Strategic Air Command (SAC) deployment plan[a]	Post–World War II demobilization; unready units; no plans, or operating concepts for rapid deployment; forces not expeditionary
Blue Bat 1958	Reassure allies; Bolster Lebanese government	SSC	Preemptive, measured	No violence; unclear how real internal threat was	Ground forces not equipped for contested assault	Existing contingency plans; forward forces and bases; expeditionary force construct	Not enough en route bases; overflight denials; base access problems;[b] lack of experience; limited range of aircraft
Second Taiwan Strait Crisis 1958	Deter invasion of Taiwan, resupply	Deter regional war	Measured	Success; no invasion of Taiwan or islands, blockade removed, no escalation	Escalation	Forward forces and bases; expeditionary force construct; prior ops in location	Other operations; lack of infrastructure; weather
Nickel Grass 1973	Resupply partner during wartime	SSC	Measured	Success; resupplied Israel with ammunition, aircraft, tanks, and other parts	N/A	Strategic airlifters; aerial refueling for C-5s	Base access problems; lacked advance plans; throughput at en route bases and aerial ports of debarkation; operational readiness of C-5 fleet; no aerial refueling for C-141
Mayaguez 1975	Recover ship and hostages	SSC	Rapid	Success; rescued crew and ship; but costly	Ground forces not prepared for contested assault	Decisive civilian leaders	Limitations on type of forces in Thailand; access problems; naval forces far away

26

Case	Purpose	Type of Operation	Response	Outcome	Risks of Rapid Employment	Enablers of Responsiveness	Hindrances
Desert Shield 1990	Deter/compel	Deter regional war	Rapid air/maritime light ground force; delayed large, heavy buildup	Success; no invasion of KSA; but did not compel withdrawal from Kuwait	Deployed forces without equipment to sustain operations; shortage of air-to-air missiles	En route infrastructure; well-developed GCC infrastructure; forward-deployed forces; aerial refueling; afloat prepositioned equipment	Initial lack of time-phased force deployment data; negotiating access; reluctance to use forces in Europe; limited APODs
Vigilant Warrior 1994	Deter aggression	Deter regional war	Measured buildup, including heavy ground forces	Success; no invasion of Kuwait		Prepositioned equipment; strategic air mobility; Civil Reserve Air Fleet; familiarity with location	Poorly packed cargo; no maintainers on prepositioned equipment; broken/inoperable equipment and vehicles in prepositioning ships
Deliberate Force 1995	Stop sectarian violence; keep coalition together	SSC	Immediate	Success; drove Bosnian Serbs to negotiations; NAC unified	N/A	Ongoing operations over Bosnia; earlier buildup of forces; aircraft on alert; forward-based forces	Access denials; political processes
Desert Fox 1998	Compel/punish	SSC	Immediate	Mixed; continued to defy inspections but degraded capability	Not having most capable forces in theater	Forward-deployed forces; forward bases	Access problems
Allied Force 1999	Stop sectarian violence; keep coalition together	SSC	Measured	Mixed; took longer than expected, killing accelerated, but eventually forced Milosevic to negotiate	N/A	Forces in theater	Inadequate infrastructure; access problems; domestic political and coalitional constraints

NOTES: GCC: Geographic Combatant Command; APOD: aerial port of debarkation.
[a] SAC war plans meant that its medium bombers were prepared for and had existing plans to deploy forward, which made them the only air forces outside of the theater able to respond within weeks to the deployment order. The Twenty-Second and Ninety-Second Bombardment Groups were operating over Korea nine days after receiving the order to deploy. In contrast, tactical aircraft based in CONUS took months to reach the theater. Robert Frank Futrell, *The United States Air Force in Korea, 1950–1953*, Washington, D.C.: Office of Air Force History, 1983, pp. 73–75.
[b] In Operation Blue Bat, U.S. forces were denied the use of neutral nations airspace and the use of some bases, which impeded the deployment of forces to the Levant. Pettyjohn and Kavanagh, 2016, pp. 97–99.

27

6. Enablers, Obstacles, and Risks of Rapid Employment

Looking across the ten cases, several key factors that enabled rapid U.S. military responses come up repeatedly. In all of the cases forward-based and deployed forces and en route infrastructure enabled U.S. forces to react in a timely fashion. This is not a surprise given that responsiveness is partially a function of distance. Proximate forces are absolutely essential for extremely short response times (minutes or hours). Given that the United States is a geographically remote power with global interests, this means that it needs to forward base or deploy forces to be highly responsive and that it also requires a robust network of en route bases and ports to facilitate the movement of troops based in the United States to hotspots around the world. Even if DoD does not have peacetime access to airbases and ports, a robust transportation infrastructure, such as was in place in Desert Shield, greatly enhances responsiveness.

If the United States has a good idea of where crises might erupt, prepositioning equipment can also dramatically improve response times, especially for heavy ground forces. If there is greater uncertainty about specifically where U.S. forces may be needed, placing prepositioned equipment on ships can improve response times from months to weeks. Other factors that have underpinned U.S. forces' ability to respond in a timely manner were the existence of contingency and deployment plans, a large strategic airlift fleet, and aerial refueling.

Conversely, the absence of these factors, particularly base access, has slowed response times.[1] In cases where there were not multiple well-developed bases that could handle large numbers of U.S. forces, bases were quickly saturated, and congestion slowed the flow of forces. This happened in Blue Bat when Wheelus Airbase unexpectedly was filled to capacity after Greece and Austria denied U.S. planes overflight rights or base access. Also during this operation, Incirlik Airbase in Turkey was the only base available in the area of operations (AO) and therefore was quickly overrun with USAF and Army forces.[2] Similarly, even when there were forward-deployed or forward-based forces, if these forces were relatively distant from the location of a contingency, it reduced the United States' ability to mount an immediate and effective response. Occasionally forces were out of position or unable to respond because they were committed to another crisis occurring at the same time. Competing demands and the sequence in which they arise, therefore, can negatively affect responsiveness. This phenomenon will be discussed further in the next chapter.

One factor that is not included in Table 5.1, but which appeared in many of the cases, is the fact that political considerations often inhibit responsiveness. The first place that politics becomes involved is in the decision to deploy forces or to intervene, which can take a good bit of

---

[1] For more on base access, see Pettyjohn and Kavanagh, 2016.

[2] Little and Burch, 1962, p. 73.

time. In several cases (Korea, Blue Bat, *Mayaguez*, and Vigilant Warrior), U.S. officials were quite decisive and opted to deploy forces in under two days. At the other extreme, there were several recent cases (Deliberate Force, Allied Force, and Desert Fox) where the United States and its allies had issued orders to deploy weeks or months in advance of the actual operation, while politicians debated about whether action was necessary. Even in some cases where U.S. forces ended up being fairly responsive (Nickel Grass, Desert Shield, and the Second Taiwan Strait Crisis), the political decisionmaking process delayed deployments by four to six critical days. For instance, Egypt and Syria attacked Israel on October 6, 1973, and Israel asked the United States to resupply it the following day. Two considerations stopped Secretary of State Henry Kissinger from immediately approving the Israeli request. First, Kissinger withheld support from Israel because it provided the United States with leverage that he needed to encourage the Israeli government to reach a cease-fire, which was the United States' immediate objective. Second, Kissinger preferred that the United States did not become directly involved in the conflict to avoid angering the Arab oil-producing states. It was only after the Soviet Union began airlifting supplies to the Arab belligerents on October 10 that Kissinger began to seriously consider the Israeli request. Still the secretary of state hesitated. It took a direct appeal from Israeli Prime Minister Golda Meir to sway President Richard Nixon, who ordered USAF to begin to airlift supplies directly to Lod Airport on October 13.[3]

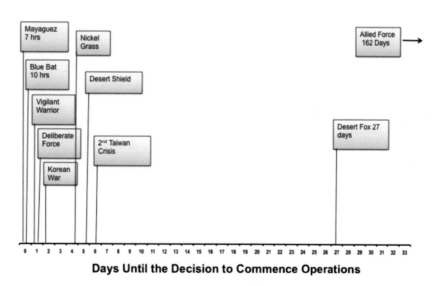

Figure 6.1. The Political Timeline to Deploy Forces in Ten Cases

The second way that politicians may inhibit effective military responses is by placing constraints on how force is used. Airmen and airpower enthusiasts often complain that airpower

---

[3] William B. Quandt, *Peace Process: American Diplomacy and the Arab-Israeli Conflict Since 1967*, 3rd ed., Washington, D.C.: Brookings Institution, 2005, pp. 107–114.

has not been employed properly because of political considerations.[4] The cases reveal that political concerns certainly shape what U.S. forces have been allowed to do. There are two different motivations for these constraints: strategic concerns and domestic political and coalitional concerns. The primary strategic concern is the risk of inadvertent escalation. This was an important factor that led the Truman administration to prohibit USAF from striking targets in Manchuria during the Korean War.[5]

Domestic and coalitional constraints may vary, but officials are often wary of taking steps that could hurt their popularity at home, such as unintentionally killing civilians or suffering a large number of military casualties. At other times, U.S. forces are constrained by the need to maintain other nations' support or by the decisionmaking process in international organizations such as NATO or the UN. This certainly played a factor in the U.S. operations in the former Yugoslavia. During Operation Deny Flight, NATO aircraft had to obtain the approval of NATO and the UN for each individual target under the so-called dual key arrangement.[6] Political sensitivities, therefore, resulted in extremely "tight control" and the "effective but inefficient use of airpower."[7] Similarly, in Allied Force, before launching an attack on a particular site, NAC and all 19 of NATO's members had to sign off on each target.[8] Not surprisingly, "combat operations by committee" introduced "inefficiencies," which frustrated U.S. military officials who believed that an intensified bombing campaign was needed to force Milosevic to capitulate.[9] In part, it seems that airpower is attractive to political leaders exactly because it can minimize the number of casualties and the risks of escalation (which have not been high since the Vietnam War).

Finally, many of the cases also highlight the fact that there are risks associated with rapidly deploying forces. During the first days of the Korean War, the only U.S. forces that were capable

---

[4] For instance, Lt Gen Michael Short, who was the air component commander during Allied Force, has been extremely critical of the gradual and highly constrained rules of engagement placed on air forces during the 1999 war in Kosovo. He wanted to "go downtown" and hit targets in Belgrade from the beginning as the U.S. had done in Iraq in 1991. For more on Kosovo complaints, see Dag Henriksen "Inflexible Response: Diplomacy, Airpower and the Kosovo Crisis, 1998–1999," *Journal of Strategic Studies*, Vol. 31, No. 6, 2008, pp. 846–850; Alan J. Vick et al., *Aerospace Operations Against Elusive Ground Targets*, Santa Monica Calif.: RAND Corporation, MR-1398-AF, 2001, pp. 14–15; Lambeth, 2001, pp. 179–182; Rebecca Grant, "Airpower Made It Work," *Air Force Magazine*, November 1999; Adam J. Herbert, "The Balkan Air War," *Air Force Magazine*, March 2009. For complaints about other operations, see Philip S. Mellinger, *Airpower Myths and Facts*, Maxwell Air Force Base, Ala.: Air University Press, December 2003, pp. 88–90; Benjamin S. Lambeth, "The U.S. Is Squandering Its Airpower," *Washington Post*, March 5, 2015.

[5] Alan Stephens, "The Air War in Korea, 1950–1953," in John Andreas Olsen, ed., *A History of Air Warfare*, Washington, D.C.: Potomac Books, 2010, p. 100.

[6] Ronald M. Reed, "Chariots of Fire: Rules of Engagement in Operation Deliberate Force," in Owen, 2000, p. 401.

[7] Reed, 2000, p. 416.

[8] After several weeks, NAC removed itself from the targeting authorization process. For more, see John E. Peters et al., *European Contributions to Operation Allied Force*, Santa Monica, Calif.: RAND Corporation, MR-1391-AF, 2001, pp. 25–29.

[9] Lambeth, 2001, p. 185.

of getting to Korea quickly were those tasked with occupation duties in Japan. An ad hoc unit—Task Force Smith—was created by hurriedly assembling 406 infantrymen and 136 artillerymen and flown to Korea on July 4. Undermanned and insufficiently equipped, the U.S. troops were quickly outflanked by North Korean forces. After six and half hours of combat, Task Force Smith had lost approximately a quarter of its men and was forced to withdraw. Most of the U.S. reinforcements that were deployed in the next ten weeks were similarly ill trained, ill equipped, and understrength and not surprisingly suffered a similar fate.[10]

In several other cases, the United States also deployed somewhat unready forces, but they were fortunate enough not to be thrust immediately into battle, where their weakness would be exposed. For instance, in Operation Blue Bat, the first battalion of Marines to land in Lebanon did not have their tanks and artillery with them.[11] If their landing had been opposed, they would have been in for a difficult fight. Similarly, in Desert Shield, U.S. decisionmakers consciously decided to prioritize getting combat power in place rather than the supporting units and equipment needed to sustain these forces.[12] Had Iraqi forces quickly moved into Saudi Arabia, it is doubtful that U.S. forces could have engaged in prolonged battle.

A course of action considered during the *Mayaguez* operation perhaps best exemplifies the risks when urgency overrides other considerations. By 1975, the Thai government prohibited the United States from stationing ground forces in its country. Consequently, there was no crisis response force on hand capable of launching a rescue operation soon after the *Mayaguez* was located. Nevertheless, because there was tremendous pressure to act quickly, one U.S. officer developed a rescue plan that relied on USAF security forces, whose mission was air base defense and thus were not trained for assault operations but were allowed in Thailand.[13] This half-baked idea was only discarded after a tragic helicopter crash killed many of the USAF security forces being transported from northern Thailand to U-Tapao to prepare for the operation.[14] If the United States had decided to go down this route because the Marines had been delayed or it was decided that U.S. forces had to strike immediately, it would have likely had disastrous results as Cambodian resistance was much stronger than expected.

---

[10] For more on Task Force Smith, see Garrett, 2000; Fehrenbach, 1963, pp. 65–71; Betts, 1995, pp. 16–17.

[11] Little, 1962, p. 19.

[12] Thomas A. Keaney and Eliot A. Cohen, *Revolution in Warfare? Air Power in the Persian Gulf*, Annapolis, Md.: Naval Institute, 1995, p. 175.

[13] This plan rested on using USAF security forces, essentially police officers who were not trained for combat operations.

[14] Wetterhahn, 2002, pp. 76–78.

7. Conclusion

The U.S. military and, in particular, USAF face a readiness problem, which DoD is seeking to fix. The demands placed on USAF by the wars in Afghanistan and Iraq coupled with shrinking force structure forced the service to defer "full-spectrum training."[1] This deficiency was exacerbated by the sequester cuts in 2013. As former USAF chief of staff General Welsh explained: "Sequestration imposed sudden and significant budget cuts and restrictions without any reduction in operational requirements while we were still fully engaged in combat operations."[2] As a consequence, USAF had to ground 31 flying squadrons, make deep cuts to training, seriously reduce pilot flying hours, furlough most of its civilian employees, and defer maintenance to aircraft, facilities, and training ranges in 2013. Although the situation has improved, it will be some time before the entire service fully reconstitutes its readiness.

The 2018 National Defense Strategy made improving the joint force's readiness a priority. Nonetheless, because the relatively high level of readiness across DoD is its single greatest expense, there are those that question whether readiness needs to be improved or whether the specter of a hollow force is one that is raised to justify increased defense expenditures. One counter to these arguments is that readiness is a critical factor that enables the U.S. military to be highly responsive, which is something that is expected of it. This study aimed to contribute to this debate on readiness by exploring the requirement for responsiveness across time.

This historical analysis demonstrates that U.S. forces have been called on to rapidly respond to crises many times since 1950 and that responsiveness was important—to varying degrees—to achieving the United States' political aims in all but one of the cases examined. Quick reactions are often needed in response to unanticipated or out-of-the blue events, such as terrorist attacks and hijackings, and they are also important when the United States is seeking to deter aggression, prevent an opponent from rapidly achieving its objectives, and assist a partner or ally that is under attack. Conversely, responsiveness tended to be less important when critical U.S. interests were not at stake but decisionmakers chose to use force as a coercive tool.

Overall, in these cases U.S. forces have proven quite responsive and typically had enough military capability in place to carry out the operation within weeks of the deployment order being issued. It was only in the instances of deterring or intervening in a major war where the U.S. required a large joint force, including heavy ground troops, that it took months to mass the requisite forces.

In the ten cases, there were many recurring factors that facilitated rapid responses, including having forward-based or deployed forces proximate to the AO, political decisiveness, prepared

---

[1] Amy McCullough, "Cutting Readiness," *Air Force Magazine*, April 2013.

[2] General Mark A. Welsh III, *Department of the Air Force Presentation to the Senate Armed Services Committee: The Impact of Sequestration on National Defense*, January 28, 2014.

en route infrastructure, prepositioned equipment, agreed-upon base access, strategic airlift, and prepared contingency and deployment plans. Not surprisingly, the absence of these factors hindered responsiveness.

Yet the past is not always a good predictor of the future security environment. As we look to the future, what range of contingencies is the joint force likely to be called on to carry out? U.S. forces have been frequently called on to respond to multiple operations at the same time, but many of these were SSCs. The Trump administration has prioritized strengthening deterrence against potential great power adversaries—in particular Russia and China—which requires a large joint force capable of rapidly responding and preventing aggressors from completing a fait accompli.

Going forward, the United States must consider the balance between readiness for smaller-scale crisis response operations and being prepared to deter and defeat a great power in major combat. During the Cold War, when the U.S. prioritized the latter, the joint force was still called on to conduct smaller operations, and simultaneity events were common. Since the end of the Cold War, the balance has shifted in favor of SSCs, and the demand for simultaneous operations has increased substantially.

It is not clear that the U.S. military can continue to try to do everything. Preparing for great power competition is likely to require U.S. policymakers to be more judicious in when they task U.S. forces with responding to other situations. The strains on the force are particularly large today because the U.S. military is mainly CONUS based. Additional presence forces are sourced through temporary rotations, which is a greater drain on service force structure compared to permanently forward-based forces. Even if U.S. forces engage in fewer SSCs, because simultaneity events happen regularly, DoD needs to consider the total possible demand for responsiveness.

Finally, how responsive is airpower? Generally, airpower has proven to be extremely responsive. In nearly all of the cases examined, 50 percent of the air forces needed were in place in three weeks or less. Moreover, unlike other types of forces, air forces offer global responsiveness. Because of their range and speed, ready air forces can quickly respond to events nearly anywhere in the world. They, therefore, do not need to be in position ahead of time to respond within days or weeks to unforeseen events around the world. Their responsiveness, however, does depend on the availability of air bases with adequate infrastructure to support operations and adequate throughput to flow forces into a distant theater. While air forces are able to quickly respond to contingencies across the globe, it is not clear that they are always the right or preferred force for a given situation. For instance, there are questions about whether airpower alone could stop a Russian short-warning invasion of the Baltics, given Russia's dense network of air defenses and how rapidly the invading force could advance. There simply might not be enough time for NATO's air forces to deliver enough weapons to stop an invading force of that size.[3] By themselves air forces also have experienced difficulties conducting rescue operations as was

---

[3] Shlapak, 2018, pp. 10–11.

evidenced in the *Mayaguez* operation. So, while airpower is highly responsive, it might not always be the capability that is required for a given situation.

This historical analysis cannot specify the exact level of readiness that the joint force or USAF should maintain. Because readiness exists along a spectrum, different readiness models and mixes of levels of readiness may be appropriate for different types of forces. But the analysis does suggest that the United States has demanded a relatively ready military since World War II and that if the American people and American policymakers continue to expect their military to be able to quickly respond to events anywhere in the world, maintaining a relatively high level of readiness will be necessary.

Appendix A: Case Studies[1]

Korean War

On June 25, 1950, over ten DPRK divisions crossed the thirty-eighth parallel and invaded ROK. The Truman administration was surprised not only by the communist aggression but also by the poor performance of the ROK military, which quickly was routed by the North Korean People's Army (NKPA). From the beginning, U.S. President Harry S. Truman was committed to preventing a DPRK victory for fear that it would lead to communist expansion across the globe and was convinced that the Soviet Union was not prepared to directly intervene in support of the north. Citing the imminent danger posed by this crisis, on June 26 Truman authorized U.S. air and naval forces to support the beleaguered ROK forces south of the thirty-eighth parallel. Four days later, the President did not consult with Congress before announcing that U.S. forces would join a UN-led police action in support of South Korea.[2] But because U.S. forces had great difficulty quickly mounting a rapid and effective defense of South Korea, it was not until September that the NKPA advance was halted.[3]

The U.S. military, which was in the midst of a massive demobilization, was not prepared to fight a major war in Northeast Asia. On paper, the United States had a sizable force in Northeast Asia, but most of these troops were understrength, prepared for occupation duties, and postured for defensive operations. At his immediate disposal, Commander of the Far East Command General Douglas MacArthur had the 8th Army's four divisions (the 7th, 24th, and 25th Infantry Divisions, and the dismounted First Cavalry Division) stationed in Japan; five fighter wings, two bomber wings, one transport wing, and miscellaneous support units (in total about 1,200 aircraft) of the Far East Air Forces (FEAF); Task Force 90 and Task Force 96 of Naval Forces Far East; and in the Philippines the 7th Fleet.[4]

The first U.S. troops able to respond to the crisis in Korea were air and naval forces based in Japan. When U.S. civilians were ordered to evacuate South Korea on June 26, it was Japan-based U.S. fighter and bomber aircraft that provided cover to the Americans leaving by sea.[5] At the

---

[1] Ideally, one would have a more detailed assessment that compares the relative demand for airpower to forces operating in other domains; however, time and data limitations precluded this type of analysis.

[2] Melvyn P. Leffler, *A Preponderance of Power: National Security, the Truman Administration, and the Cold War*, Stanford, Calif.: Stanford University Press, 1994, pp. 364–368; Field, 1962, chap. 3.

[3] Alan R. Millett and Peter Maslowski, *For the Common Defense: A Military History of the United States of America*, New York: Free Press, 1994, p. 510.

[4] Task Force 90 was the core amphibious task force, while Task Force 96 included one cruiser, four destroyers, some minesweepers, and other small vessels, but none of its ships had anything larger than a 5-in gun. Field, 1962, chap. 3.

[5] Futrell, 1983, p. 8; Warnock, 2000, p. 2.

same time, the 7th Fleet's USS *Valley Forge* Battle Group began to sail north, and FEAF repostured its aircraft so that they were stationed at bases that put them within range of South Korea. The next day, transport aircraft from the 374th Troop Carrier Wing evacuated another 750 Americans from Korea to Japan and were supported by 5th Air Force F-82s, which engaged in the first air combat of the war.[6]

The United States had managed to get most of its citizens off the peninsula but had done little to slow the NKPA, and on June 28 the capital of South Korea, Seoul, fell. That same day USAF flew its first strike missions of the war. FEAF B-26 and B-29 bombers begin flying interdiction missions against railroads and columns of moving North Korean forces on the roads. On the twenty-ninth, FEAF bombers were authorized to strike North Korean airfields.[7] When the *Valley Forge* got in position off DPRK on July 3, its Corsair and Skyraider aircraft also began to bomb North Korean targets.[8] U.S. air and sea power, however, proved to be too little and too late to stop the invasion.

But air and sea forces were the only forces able to respond because the United States had no ground combat forces in Korea and the movement of ground forces from Japan to Korea proceeded slowly. The entire 24th Division had been ordered to deploy to Korea on June 30, but its movement was delayed by the lack of sealift and the fact that the division was dispersed across the Japanese islands.[9] The first U.S. ground unit to arrive in Korea on July 1 was the ill-fated Task Force Smith, an undersized infantry unit, consisting of only 406 soldiers.[10] The size and composition of Task Force Smith had been determined by the number of C-54 transport aircraft that were immediately available in Japan to lift the soldiers to South Korea.[11] These U.S. soldiers that were armed with only rifles, mortars, and rocket launchers were tasked with delaying the advancing North Korean armies near Osan. Undermanned and insufficiently equipped, the U.S. troops were quickly outflanked by heavy North Korean forces. After six and a half hours of combat, Task Force Smith had lost approximately a quarter of its men and was forced to withdraw.

Additional reinforcements from Japan, which were traveling by sea, took days, if not weeks, to arrive. The remainder of the 24th Infantry Division did not close in Korea until July 6 with the

---

[6] Warnock, 2000, p. 2.

[7] Warnock, 2000, p. 3.

[8] Field, 1962.

[9] Fehrenbach, 1963, p. 65.

[10] Task Force Smith was ordered to confront and delay the advancing North Korean forces to buy time for additional American reinforcements to arrive, but it was inadequate for this mission and was routed in several hours. Task Force Smith consisted of two understrength infantry companies along with some headquarters and communications personnel, totaling 406 soldiers. They were armed with their rifles and two 75-mm recoilless rifles, two 4.2-in mortars, six 2.36-in rocket launchers, and four 60-mm mortars. They were later joined by part of an artillery unit with six 105-mm howitzers. Appleman, 1992, p. 61; Fehrenbach, 1963, pp. 65–66.

[11] Garrett, 2000, p. 6.

25th Infantry Division hot on its heels.[12] The 25th Infantry Division began to embark on ships in Japan on July 8, and by the twenty-fifth of July, the final elements of the division had arrived in South Korea.[13] The 1st Cavalry Division began to arrive at Pohangdong, South Korea, on July 18 and closed within two days.[14] The final intratheater movement of ground forces occurred on July 24 when two battalions of the 29th Infantry Regiment based on Okinawa landed at Pusan.[15] Yet nearly all of the U.S. ground forces that arrived in these early days of the war were ill trained, ill equipped, and understrength and not surprisingly did not do much to turn the tide of the war.[16] The 8th Army, under the command of Lieutenant General Walton H. Walker, struggled to even hold a defensive perimeter around the city of Pusan and prevent South Korea from collapsing entirely, while waiting for reinforcements from the United States to arrive.[17]

In part, the failure of U.S. forces on the ground was due to problems that U.S. air forces experienced in the air war. FEAF was equipped for a limited defense of Japan and the Philippines, not an offensive war over Korea. Only approximately half of its assigned aircraft were operational, and its most numerous aircraft, the F-80 Shooting Star, proved ill suited for this war. F-80s had a limited range and could only remain on station over Korea for a relatively short period. Moreover, the Shooting Stars required 7,000-foot runways to operate, of which there were only four in Japan and none in South Korea.[18] Moreover, the F-80s were outmatched by the faster Soviet-built MiG-15s flown by the NKPA air force. Consequently, in the first week of July, FEAF decided to pull out of storage F-51 Mustangs, an older, piston-engine aircraft that had a longer range and could fly from austere airfields for the ground attack mission. This process of collecting F-51 aircraft, which were mainly in the guard and reserve, took time. On July 24, 145 F-51s arrived on the USS *Boxer* and quickly were flying missions against the NKPA forces.[19]

It was not until September that the United States had assembled a joint force that was capable of going on the offensive against the North Koreans. In the first month of the war, few U.S. reinforcements arrived from outside of the East Asian theater beyond the two bomber groups that were deployed from CONUS to Japan.[20] Beyond the 7th Fleet, there were few additional maritime reinforcement available as most of the Pacific fleet was stationed on the West Coast of

---

[12] Garrett, 2000, p. 7.

[13] Gordon Rotterman, *Korean War Order of Battle: United States, United Nations, and Communist Ground, Naval, and Air Forces, 1950–1953*, Westport, Conn.: Praeger, 2002, p. 29.

[14] Rotterman, 2002, p. 21.

[15] Field, 1962.

[16] For more on Task Force Smith, see Garrett, 2000; Fehrenbach, 1963, pp. 65–71; Betts, 1995, pp. 16–17.

[17] The 8th Army had less than one-fifth of its authorized tanks. Millett and Maslowski, 1994, p. 510.

[18] Futrell, 1983, pp. 59–60.

[19] Warnock, 2000, p. 5.

[20] Futrell, 1983, pp. 73–74.

the United States.[21] Additional ground forces were the last to arrive. The 1st Marine Provisional Brigade was activated at Camp Pendleton July 7 and set sail a mere five days later. It took the Marine brigade a little over a month before it closed in South Korea on August 17. The first elements of the 2nd Infantry Division left Washington by sea for Japan on July 17 and arrived in Korea on July 31. The division's deployment, however, was not completed until August 20.[22]

On July 25, the Joint Chiefs of Staff (JCS) directed USMC to build up the 1st Marine Division to full war strength and deploy it to Korea. The final elements of the 1st Marine Division joined up with the 1st Provisional Marine Brigade already in Korea and unloaded at Incheon on September 21.[23] In late August elements of the 3rd Infantry Division and 65th Infantry Regiment left from San Francisco and Puerto Rico bound for Korea. The last of the 3rd Division units closed on September 15, while the last of the 65th Infantry units did not arrive until October 1 due to the fact that one of its transport ships broke down en route.[24]

In mid-September, General MacArthur launched the daring amphibious landing at Incheon that turned the tide of the war. As DPRK reeled from the surprise, the 8th Army went on the offensive, driving the NKPA back toward the thirty-eighth parallel. The ground offensive was supported by a major B-29 strategic bombing campaign and the movement of FEAF fighter squadrons from Japan to Korea, where they provided vital air support to the U.S. offensive.[25] In sum, it took the United States nearly three months to build up and deploy sufficient forces to repel the DPRK invasion of South Korea.

Operation Blue Bat

In July 1958, President Dwight D. Eisenhower ordered U.S. forces to intervene in support of the government of Lebanon. As a result, within 19 days the United States deployed 55 combat aircraft, 87 ships, and nearly 15,000 ground forces to the small Middle Eastern nation. This operation was in response to the July 14 request for U.S. military support by Lebanese President Camille Chamoun, who invoked the Eisenhower doctrine—the U.S. president's commitment to support anticommunist forces in the Middle East. It appeared as if the pro-Western Lebanese government might be on the brink of falling as a result of a crisis that had been precipitated due to Chamoun's efforts to amend the constitution so that he could serve another term in office. There was little evidence that Chamoun's opponents were communists or had links to the Soviet

---

[21] Field, 1962. The first aircraft carrier, the USS *Boxer*, did not dock at Yokosuka until July 23. Warnock, 2000, p. 5.

[22] Gough, 1987, pp. 4–6.

[23] Donald K. Wols, "Relationship Between Cohesion and Casualty Rates: 1st Marine Division and 7th Infantry Division at Inchon and the Chosin Reservoir," thesis, Fort Leavenworth, Kan.: U.S. Army Command and General Staff College, 2004, p. 28.

[24] Gough, 1987, pp. 6–8.

[25] Warnock, 2000, p. 9.

Union. Nevertheless, pro-Nasserist military officers had just deposed the pro-Western Hashemite monarchy in Iraq on July 14. The Eisenhower administration feared that inaction would harm the United States' reputation and that its access to bases in the region would be at risk.[26]

For more than 6 months, Lebanon had been teetering on the brink of a civil war; consequently, DoD had been seriously planning an intervention since November 1957.[27] A large U.S. naval amphibious force carrying three battalions of Marines had been ordered to move toward Lebanon on May 16. But on July 15, only one of the three amphibious task forces was close enough to Lebanon to immediately land. The first U.S. Marines went ashore four miles south of Beirut near the airport in Lebanon, 26 hours 20 minutes after Chamoun requested U.S. assistance. The initial landing team lacked artillery and tanks, which were on a ship farther away, and could have presented a problem if the Marines had encountered resistance.[28] Air support was provided by carrier aircraft flying over 700 miles each roundtrip and therefore staging through Cyprus.[29] Because of the lack of resistance, in just over an hour the Marines reached the Beirut airport.

Also on the morning of July 15, Tactical Air Command was ordered to deploy its CASF Bravo instead of U.S. Air Forces in Europe deploying forces as had originally been planned.[30] The leading elements of the CASF—the 12 B-57s, 12 F-100s, and RF-101s—left their east coast bases by midafternoon. Two additional waves of aircraft left the following day, while the second Marine battalion began to land in Lebanon. The first F-100s landed at Incirlik Airbase in Turkey on July 16 after flying 12.5 hours with three aerial refuelings. By the evening of July 17, 15 F-100s and 10 B-57s were on station in Turkey. On July 18, the third battalion of Marines landed at a beach north of Beirut, and a fourth battalion deployed from CONUS and airlifted from Morocco began to arrive and were held as an afloat reserve.

On July 19, U.S. Air Forces in Europe aircraft began to transport paratroopers from the 24th Infantry Division's 19th Airborne Battle Group from Germany to Beirut with a two-day stopover in Incirlik.[31] On July 20 the CASF deployment had closed.[32] The Army unit closed in Lebanon on July 26 with flights being delayed by base access and overflight restrictions. Additional soldiers, tanks, artillery, and support materials were embarked on ships in France and Germany and arrived in Lebanon in early August.

---

[26] Byron R. Fairchild and Walter S. Poole, *The Joint Chiefs of Staff and National Policy, 1957–1960*, Vol. 7, Washington, D.C.: Office of Joint History, 2000, p. 155.

[27] Fairchild and Poole, 2000, p. 153.

[28] Dragnich, 1970, p. 34.

[29] Dragnich, 1970, p. 33.

[30] CASF Bravo consisted of two squadrons of F-100D/F fighters, one squadron of B-57 tactical bombers, one squadron of KB-50 tankers, and one squadron of mixed reconnaissance and command and control aircraft. Little and Burch, 1962, p. 29.

[31] Little and Burch, 1962, pp. 24–25.

[32] Little and Burch, 1962, pp. 30–31.

In the end, this relatively large force was deployed to Lebanon but did not have to fight. It proved to only be a show-of-force operation aimed at deterring antigovernment Lebanese forces. The crisis was resolved with U.S. mediation. President Chamoun was allowed to complete his term, and a moderate was elected to replace him.

Second Taiwan Strait Crisis

Tensions started to rise over the Taiwan Strait in July 1958.[33] Things came to a head on August 23 when Chinese communist forces launched 40,000 artillery shells at the Quemoy Islands, which are located less than five miles from the shore of China and where the Taiwanese had garrisoned more than a 100,000 soldiers. After the initial barrage, the Chinese used patrol torpedo boats and indirect fire to blockade the Quemoy Islands, cutting off the Taiwanese troops from crucial supplies. Although U.S. officials did not believe that an invasion was imminent, on August 25 President Eisenhower ordered the reinforcement of U.S. air defenses on Taiwan, the augmentation of the USN's 7th Fleet, and preparations to begin to escort Taiwanese resupply ships. The latter measure was not publicly announced for several days because the Eisenhower administration did not want to embolden the Taiwanese government, which U.S. officials feared was trying to draw the United States into a war with China. Therefore, while the United States sought to deter a Chinese invasion of the Quemoy and to break the blockade of the offshore islands, it also wanted to restrain the Taiwanese and keep the conflict limited.

On August 24 and 25, the carriers *Essex* and *Midway* left the Mediterranean and Hawaii, respectively, and sailed for Taiwan.[34] It was not until the twenty-ninth of August that the CASF based in CONUS and a fighter squadron based on the Japanese island of Okinawa were ordered to deploy because of the ongoing situation in Lebanon and uncertainties about Chinese intentions.[35] All but two planes from the Japanese-based 16th Fighter-Interceptor Squadron were in position and combat ready within 7.5 hours of the order to deploy being issued. After a 10-day transit, the *Midway* arrived near Taiwan on September 4. Fighter and reconnaissance aircraft that were a part of a CASF were able to self-deploy with the aid of aerial refueling within 2 to 6 days of their departure. These transits were longer than usual due to storms and maintenance issues. The F-104 Starfighters, which were tasked with air defense, however, had to be disassembled and transported on C-124 transport aircraft, which took even longer. The last F-104 did not arrive until September 19, closing the CASF deployment. Sea-based aircraft took nearly as long to arrive because the *Essex* CVBG had to sail 22 days and transit the Suez Canal, entering the AO on September 16.

---

[33] For background on the crisis, see Halperin, 1966.

[34] Van Staaveren, 1962, p. 23.

[35] Van Staaveren, 1962, p. 19.

Aircraft from Marine Air Group 11 stationed in Atsugi, Japan, were alerted on August 24, but they did not begin to deploy for more than two weeks because of the lack of airfields that met Marine requirements. The 56 aircraft from Marine Air Group 11 were finally on Taiwan on September 18.[36] JCS instructed the Army to deploy a Nike-Hercules battalion of surface-to-air missiles from Texas to Taiwan on August 25, but it did not even begin to deploy from Fort Bliss for over a month. During this interval, a battalion headquarters and four firing sites on Taiwan were selected, and construction of the required semipermanent facilities needed to support the battalion began on September 15. The leading edge of the 71st Artillery 2nd Missile Battalion arrived in Taiwan by airlift on September 17, and the rest of the unit arrived by sea on October 9. The Nike battalion was not operational until October 25.

The U.S. military response compelled China to end the blockade and deterred Beijing from invading the Quemoy Islands. The situation had already deescalated by early September when Chinese bombardments had tapered off, and on September 6 China offered to reopen ambassadorial talks with the United States.[37] On October 5, the Chinese government proposed a cease-fire and an end to the blockade conditioned on the United States halting convoy operations. President Eisenhower accepted the offer and induced the Taiwanese to go along by promising them additional U.S. weapons in return for reducing the number of forces on the Quemoy Islands.[38] The cease-fire was broken several times and resulted in a tacit agreement in which the Chinese and Taiwanese only bombed each other every other day, which continued for nearly twenty years.

Operation Nickel Grass

On October 6, 1973, during the Jewish holiday of Yom Kippur, Egyptian forces crossed the Suez Canal and overran Israeli defenses in the Sinai, while Syrian troops attacked the Israeli-controlled Golan Heights. Although there were warnings that the Arabs were planning to attack, the Israeli and U.S. governments largely disregarded them as they did not expect the weaker Arab forces to go on the offensive. The advancing Arab armies made gains on both fronts against the understrength IDF, which had only started a partial mobilization two days before the war began. On October 7, Syrian forces made a breakthrough on the Golan Heights with the Israeli air forces suffering substantial losses. The same day, the Israeli embassy urgently requested several hundred additional Sidewinder air-to-air missiles, as well as 300 M-60A1 tanks, 40 F-3s, and jammers to use against Egyptian air defenses.[39]

---

[36] Van Staaveren, 1962, pp. 24–25.

[37] Halperin, 1966.

[38] Fairchild and Poole, 2000, pp. 213–216.

[39] Walter S. Poole, *The Joint Chiefs of Staff and National Policy, 1973–1976*, Washington, D.C.: Office of Joint History, 2015, p. 169.

Some within the Nixon administration were hesitant to fulfill Israel's request for fear of provoking an Arab oil embargo. Secretary of Defense James Schlesinger explained, "Our shipping any stuff into Israel blows any image we may have as an honest broker."[40] Moreover, at this time, most U.S. officials, including the intelligence community, expected a quick and decisive IDF counterattack, so immediate assistance was not believed to be necessary. Some U.S. policymakers also were skeptical of Israel's claims about its losses and its supply levels. Nevertheless, on October 9, President Nixon decided to approve Israel's request for resupply, including tanks, F-4 fighters, and ammunition, but he wanted to keep the American military support out of the public eye; therefore, the initial shipments were given to the Israelis, who were responsible for transporting them to their country in unmarked El Al aircraft.[41] Transferring the equipment by sea took at least 12 to 14 days from the East Coast of the United States, which was deemed to be too slow, so airlift was required. The United States also unsuccessfully attempted to charter commercial aircraft to carry shipments to Israel. Despite the lack of external assistance, on October 9, IDF turned the tide in the Golan Heights and forced the Syrians to retreat.

These indirect methods of delivery were inadequate to meet Israel's growing demands for military equipment. On October 10, 70 Soviet aircraft began to airlift supplies to Egypt and Syria, raising the stakes for the United States, which did not want to be seen as being upped by the Soviets.[42] Dissatisfied with the plans for resupply outlined by Secretary Schlesinger, Israeli Ambassador Simcha Dinitz delivered a message from Israeli Prime Minister Golda Meir to President Nixon claiming that IDF would be out of ammunition in two to three days on October 12.

The following day, the president authorized an overt airlift to resupply Israel. Suspecting that DoD was dragging its feet, Secretary of State Henry Kissinger demanded there be no further delays in providing Israel with equipment. On October 13, at 2055, the first three C-5s loaded with ammunition left Dover Air Force Base and arrived the next day at Lod Airport in Tel Aviv. Some of the equipment in this first load was reportedly delivered to the Israeli front lines within three hours of being off-loaded at Lod Airport.[43] Also on October 14, Egyptian heavy forces that had moved outside of their air defense umbrella suffered heavy losses, setting the stage for the Israelis to launch a counterattack between two Egyptian armies across the Suez Canal the next day.

MAC was directed to complete 4 C-5 and 12 C-141 deliveries a day. Once the airlift began in earnest, MAC quickly boosted Israelis military supplies, despite the long distances to be traveled. On average, MAC bases on the East Coast of the United States were 6,450 miles from Tel

---

[40] Quoted in Poole, 2015, p. 169.

[41] Poole, 2015, p. 170.

[42] William B. Quandt, *Soviet Policy in the October 1973 War*, Santa Monica, Calif.: RAND Corporation, R-1864-ISA, May 1976, pp. 19–20.

[43] Patchin, 1974, p. 14.

Aviv.[44] Within six days, the transport aircraft had completed more than a hundred missions, and after ten days, the United States had delivered more than 200 cargo loads of aircraft to Israel.[45] The airlift itself was hindered by the fact that U.S. aircraft only had access to Lajes Airbase on the Portuguese Azores, and the Portuguese government only consented to allow the United States to use Lajes Airbase in the Azores (but not airbases on the Portuguese mainland) under heavy U.S. pressure.[46] Over the course of the operation, MAC aircraft flew 567 sorties, carrying 22,328 tons of military supplies and equipment to Israel.[47]

Initially, the United States did not respond rapidly to Israel's requests for military assistance, but this was a political delay—not an operational one. While wanting to support Israel and ensure that it prevailed in the war, the Nixon administration also sought to appear neutral so as not to offend the Arab states. Israel had requested U.S. military resupply the day after Egypt and Syria attacked, but in the early days of the war, President Nixon was only willing to covertly provide military assistance to Israel. Six days passed before the United States reversed course and ordered U.S. military transport aircraft to airlift the supplies that Tel Aviv had requested.

In the end, Israeli supplies were replenished, and IDF was able to repel the Syrian and Egyptian offensives, leading to the first real Arab-Israeli peace negotiations. It is not clear how important the U.S. supplies were to turning the tide in favor of Israel. The Israelis had already begun to shift the battle in their favor before the airlift could take effect. One could argue that knowing that additional supplies were on the way freed the Israeli leadership to take the offensive and not worry about conserving their limited stocks of weapons and munitions. Supporting Israel, however, came with a cost. In retaliation, the Arab states embargoed oil against the United States and, more importantly, also cut oil production, causing a huge leap in prices in the United States.[48]

Mayaguez Rescue

On May 12, 1975, a Khmer Rouge patrol boat commandeered a U.S. civilian merchant ship, the SS *Mayaguez*, as it sailed from Hong Kong to Thailand, in waters that were claimed by the Cambodian government.[49] The Cambodians quickly seized the *Mayaguez*'s 40-man crew, but not before one of the crew members was able to send a distress message that was received by an American civilian in Indonesia, which was relayed from the U.S. embassy in Jakarta to the U.S. president. Approximately four hours after the ship had been seized, JCS ordered Pacific

---

[44] Patchin, 1974, p. 9.

[45] Patchin, 1974, p. 178.

[46] Interview with William B. Quandt, May 29, 2013.

[47] Patchin, 1974, p. 178.

[48] Poole, 2015, pp. 168–173.

[49] All dates and times for the *Mayaguez* rescue are reported in terms of Cambodian time, GMT +7.

Command to deploy reconnaissance aircraft to find the ship. Less than an hour and a half after receiving the order, a P-3 Orion aircraft took off from Thailand and began to search for the *Mayaguez*.[50] Two factors seem to have prompted U.S. President Gerald Ford to rapidly order military forces to find and rescue the ship and its crew: the memory of the *Pueblo* seizure, and the desire for the United States to appear strong and decisive just a month after communists had seized control of South Vietnam and Cambodia.[51]

Although the United States had withdrawn from the Vietnam War, the U.S. military still had a robust presence in Thailand, which included 2 tactical fighter wings, a special operations wing, a P-3 detachment, and an aerospace rescue and recovery squadron, but no ground forces.[52] The closest rapid response forces was either the battalion of Marines based in Okinawa or the Marine battalion stationed in the Philippines. Additionally, the USN 7th Fleet was based in the Philippines and included the *Coral Sea* CVBG, an amphibious assault ship, several destroyers, and a World War II era *Essex* class carrier. The *Coral Sea* and the destroyers steamed for the Gulf of Thailand, but given their starting locations, none of the ships would be able to reach the area for several days.

The White House ordered U.S. maritime patrol aircraft stationed at Cubi Point in the Philippines and at U-Tapao in Thailand to search for the missing merchant ship. After searching the Gulf of Thailand for most of the day, the morning of May 13 a P-3 positively identified the *Mayaguez*, which was sailing northeast and eventually stopped at the island of Koh Tang.[53] Throughout the day, Thai-based USAF F-4s, A-7s, F-111s, and AC-130s monitored the situation and attempted to stop the Cambodians from taking the *Mayaguez* crew to the mainland by attacking ships that approached the island.

A plan was developed to rescue the crew and liberate the *Mayaguez* with a simultaneous assault on the ship and the island of Koh Tang. Unbeknownst to the U.S. military planners, the crew already had been moved to a fishing vessel. On the fourteenth of May, USAF CH-53 Knife and HH-53 Jolly Green Giant helicopters flew from their base at Nakhom Phanom to U-Tapao, the closest Thai base to the Koh Tang and the location where U.S. forces were staged for the rescue. On the morning of the fourteenth, two Marine companies transported by MAC C-141s arrived from the Philippines at U-Tapao, while the Marine battalion landing team from Okinawa did not reach U-Tapao until midafternoon.[54]

President Ford ordered the Marines to launch an operation to recover the *Mayaguez* crew at around four in the morning on May 15.[55] The rescue operation began ten minutes later when

---

[50] Wetterhahn, 2002, p. 334.

[51] Haulman, 2000a, p. 106; Wetterhahn, 2002, p. 38.

[52] For the USAF forces in theater, see Chun, 2011, p. 20; Haulman, 2000a, pp. 105–114; and Wetterhahn, 2002.

[53] Wetterhahn, 2002, p. 334.

[54] Haulman, 2000a, p. 108; Wetterhahn, 2002, pp. 336–337.

[55] Wetterhahn, 2002, p. 338.

three USAF helicopters transported 48 Marines to the *Holt* destroyer, which had just arrived in the theater and was sailing toward the *Mayaguez*. The Marines boarded the hijacked ship only to find that it was empty. Believing the crew was on Koh Tang, it was decided to proceed with the next stage of the operation.

Eight helicopters carrying around 180 Marines attempted to land on Koh Tang beginning at 6:20 a.m. Instead of several dozen lightly armed defenders, the U.S. assault encountered a Cambodian battalion armed with rocket launchers, mortars, and heavy caliber machine guns, which fired on the USAF helicopters as they approached. Three helicopters were shot down, while most others were heavily damaged, with only one escaping mostly unharmed. Once on the ground, the Marines were engaged in a tough firefight and were outnumbered and outgunned. Also in the morning, the *Coral Sea* arrived in the theater and its aircraft began to launch attacks against targets on the Cambodian mainland to prevent the Khmer Rouge from reinforcing Koh Tang.

Meanwhile a USAF reconnaissance aircraft located the *Mayaguez* crew on the fishing boat, and the *Wilson* guided-missile destroyer (DDG) rescued them at 9:49 a.m.[56] Even though the crew had been recovered, the battle on Koh Tang continued. Marine reinforcements were transported to the island midday, and USAF provided additional air support to the fight, but the battle remained inconclusive. At around 6:00 p.m., the three remaining operable helicopters incrementally began to withdraw the remaining 200 Marines, and the last U.S. forces departed the island at 6:40 p.m. In the course of the hasty withdrawal, it was discovered that three marines had been left behind. The rescue operation had succeeded at freeing the *Mayaguez* and its crew, but the price was high. During the 14-hour battle, there were 91 U.S. casualties in the operation, four helicopters were destroyed, and eight others were heavily damaged.[57]

Forward-based forces enabled the United States to rapidly find the *Mayaguez* and within several days to launch a rescue operation. Nevertheless, because of a prohibition on U.S. ground forces in Thailand, it was nearly 62 hours before the Marines arrived in theater and were ready to launch the rescue operation. Moreover, because key naval forces, including the aircraft carrier USS *Coral Sea*, could not arrive early enough, the United States chose to ignore the Thai government's prohibition on U.S. forces based in Thailand participating in the rescue operation, which significantly damaged the relationship between the two nations. In short, U.S. forces were quite responsive, but this came at some cost politically and harmed U.S.-Thai relations.

Time was of the essence in the rescue operation. From the *Pueblo* seizure, the United States had learned that the minutes and hours after a vessel was seized were critical for getting it back. In this instance, the United States was focused on rescuing the crew before they were moved to the Cambodian mainland and likely out of reach. The pressures of the situation and the desire to act quickly led U.S. forces to make some errors and to consider some very risky courses of

---

[56] Wetterhahn, 2002, p. 339.

[57] Haulman, 2000a, pp. 109–112.

action. For instance, the absence of a rapid response force led to the consideration of a plan using USAF security forces stationed in Thailand. The USAF security forces were not trained for assault operations but instead were prepared for airbase defense.[58] Nevertheless, the U.S. began to take steps to execute this plan, which included airlifting security forces from northern Thai bases to U-Tapao, and abandoned this course of action only after a Knife helicopter carrying security forces crashed on its way to U-Tapao.[59] If the security forces had attempted to execute the assault on Koh Tang, they probably would have been slaughtered by the heavily armed Cambodian forces on the island. Time pressures also led U.S. forces to begin the rescue operation without adequate intelligence on the location of the crew and on the Khmer Rouge defenses on Koh Tang. As a result, the Marines were expecting light resistance and to find the *Mayaguez* crew on the island, when in fact they had needlessly flown into a costly ambush.[60]

Operation Desert Shield

On August 2, 1990, Iraqi forces rolled across the border and quickly overran Kuwait. In response to this action, on August 7, the United States commenced Operation Desert Shield, a large deployment of air, ground, and naval forces to the Middle East to deter further Iraqi aggression and defend KSA should deterrence fail.[61] General Hansford T. Johnson, the commander of U.S. Transportation Command, noted that in many ways "Desert Shield was a worst case scenario" because "warning time was extremely limited," there were "virtually no forces in place," and DoD had to "move our forces a great distance." Nevertheless, DoD successfully had naval, air, and light ground forces in theater within days of the deployment order being given and over the next several months built a large deterrent force that included heavy ground forces.[62]

At the time of the invasion, the United States had relatively few combat forces in the region that could effectively defend KSA should it be attacked. The longstanding USN Middle East Force had 6 small ships; there were 14 F-111Es stationed at Incirlik Airbase in Turkey and 2 KC-135 tanker aircraft in UAE for an exercise. Two aircraft carriers were nearby, but not immediately within range of conducting air operations over KSA. The *Independence* CVBG was in the Indian Ocean, while the *Eisenhower* CVBG was in the Adriatic. In sum, a limited amount

---

[58] This plan rested on using USAF security forces, essentially police officers who were not trained for combat operations.

[59] Wetterhahn, 2002, pp. 76–78.

[60] Haulman, 2000a, pp. 111–112.

[61] The forces used for Operation Desert Shield were later employed to liberate Kuwait, but that decision was not made until late October and implemented in the phase 2 deployment, which is not considered here. As the United States initiated operations against Iraqi forces at the time and place of its choosing, responsiveness was not particularly important.

[62] Quoted in William R. Tefteller, *Strategic Airlift Support for U.S. Forces Deployment to Operation Desert Shield*, Washington, D.C.: National Defense University, 1991, p. 1.

of U.S. air and maritime forces was able to respond within days to an attack, but there were no ground forces in the AO, and there was no heavy ground equipment, such as tanks and armored fighting vehicles, in place, which would have been needed to repel an invading armored force.

In an effort to rectify this situation, the United States quickly moved a joint force into the theater as a deterrent to further Iraqi aggression. To expedite the arrival of a credible deterrent, Commander-in-Chief U.S. Central Command decided to delay the deployment of logistics and support forces in favor of combat forces. This meant that many Army ground combat units only had their organic supplies and supporting assets and that once these were exhausted, they had to rely on host nation support and USMC supplies of food and water.[63] Similarly, additional supplies of USAF munitions were deferred, leaving the combat aircraft in theater to rely on stocks of prepositioned munitions that could have only supported operations for a short period.[64] On August 7, deployment orders were issued for the ready brigade of the 82nd Airborne Division, maritime prepositioning squadrons in Guam and Diego Garcia, the *Eisenhower* CVBG, the USS *Wisconsin* battleship, and 2 F-15C squadrons, along with enough E-3 Airborne Warning and Control System (AWACS) and tanker aircraft to put in place one 24-hour combat air patrol.[65]

Nearby maritime forces were the first to arrive in the theater. First, the aircraft carrier USS *Independence*, which had been sailing toward the region since August 4, arrived on station in the Gulf of Oman, while the USS *Eisenhower* CVBG crossed through the Suez Canal, entering the theater.[66] The following day (C+1) the first forces from CONUS, 23 F-15Cs from the 71st Tactical Fighter Squadron and 5 E-3s, landed at Dhahran Airbase and immediately began defensive combat air patrols.[67] On August 9 (C+2), a second squadron of F-15C aircraft touched down in Saudi Arabia, and the first elements of the 82nd Airborne Division also landed in the kingdom and secured the perimeter around Dhahran Airport.[68] By the tenth (C+3), additional air forces arrived in the theater, including 24 F-16Cs at Al Dhafra Airbase in UAE, 25 F-15Es at Thumrait (Oman), 12 British Royal Air Force Tornado F-3s, and 2 RC-135 Rivet Joints.[69] Yet

---

[63] DoD, 1992, p. 43.

[64] Kiraly, Owen, and Pinker, 1993, p. 16.

[65] For a detailed description of every unit activated and its deployment, see Kiraly, Owen, and Pinker, 1993; Edward J. Marolda, *The United States Navy and the Persian Gulf War*, Naval History and Heritage Command (website), August 23, 2017; Robert H. Scales, *Certain Victory: The US Army in the Gulf War*, Fort Leavenworth, Kan.: U.S. Army Command and General Staff College Press, 1993, appendix; James K. Matthews and Cora J. Holt, *So Many, So Much, So Far, So Fast: United States Transportation Command and Strategic Deployment for Operation Desert Shield/Desert Storm*, Washington, D.C.: Joint History Office, 1992.

[66] Kiraly, Owen, and Pinker, 1993, p. 10.

[67] The 71st Tactical Fighter Squadron began to deploy 18 hours after receiving the order. Their deployment required seven aerial refueling. Richard G. Davis, *On Target: Organizing and Executing the Strategic Air Campaign Against Iraq*, Washington, D.C.: U.S. Air Force, 2002, p. 40; Kiraly, Owen, and Pinker, 1993, p. 12.

[68] Davis, 2002, p. 41.

[69] Kiraly, Owen, and Pinker, 1993, p. 16.

these combat air forces had no real air-to-ground attack capability because of "malpositioned" munitions.[70]

The first fast sealift shift intended to carry the 24th Infantry Division (Mechanized) arrived in Savannah, Georgia, on August 10 (C+7), and three days later it embarked for KSA.[71] The first maritime prepositioning ship arrived in Saudi Arabia on August 15. A little over a week after the deployment order was issued (C+8), the entire 2nd Brigade of the 82nd Airborne was in place, 245 USAF aircraft were in the theater (including 134 fighters), and advance elements of the 7th MEB had disembarked at Jubayl, Saudi Arabia.[72]

Although the United States quickly deployed light ground-based units to Saudi Arabia, this force was little more than a trip wire. It was not until August 27 that the first M1 Abrams tank was off-loaded from a fast sealift ship in Saudi Arabia.[73] Heavy ground forces did not arrive en masse for at least another 30 days.[74] By contrast, only 22 days after receiving its order to deploy, the light 82nd Airborne Division was able to close in theater.[75] Similarly, USN units that had to deploy from CONUS or other distant theaters were not on station for several weeks, if not longer.[76] By contrast, USAF aircraft continued to rapidly arrive in the Persian Gulf (see Figure 2.4). Within 16 days, 384 aircraft, which was half of the force allocated for the initial deployment (through November), had arrived in the Persian Gulf.

The forces deployed as a part of Operation Desert Shield were sufficient to dissuade Saddam Hussein from attacking Saudi Arabia, although it is not known if he intended to do so. The operation was less successful at a secondary objective that emerged over time—compelling Hussein to withdraw from Kuwait.[77] The United States' immediate response to Iraqi aggression was limited to light ground forces, sea- and land-based airpower, and maritime forces. If Hussein had decided to attack the U.S. trip wire force, he may have initially succeeded because of the lack of heavy ground forces and the fact that many of the first U.S. forces deployed lacked important support functions and critical equipment, like munitions, that would have been needed if they were to have engaged in combat operations. The United States was able to rapidly deploy forces to this far-off region where it had few permanently based forces because of the existence

---

[70] I infer that this means that they were armed with air-to-air munitions and that no stocks of air-to-ground munitions were readily available. Kiraly, Owen, and Pinker, 1993, p. 17.

[71] Scales, 1993, p. 391.

[72] Kiraly, Owen, and Pinker, 1993, pp. 9–21.

[73] Scales, 1993, p. 392.

[74] The 1st Cavalry Division closed in theater on October 22 (C+76). Scales, 1993, p. 392.

[75] Scales, 1993, p. 392.

[76] For instance, other major naval combatants, including the *Saratoga* and *Wisconsin*, arrived in late August, while the ARG carrying 2/4 MEB and *Kennedy* not until mid-September. DoD, 1992, p. E-24.

[77] Janice Gross Stein, 1992.

of a robust en route transportation infrastructure in Europe and because it was given access to the many capable airbases in the Persian Gulf region.

Operation Vigilant Warrior

On October 5, 1994, U.S. intelligence agencies reported that two elite armored divisions of the Iraqi Republic Guard were moving from near Baghdad toward the border with Kuwait. Within seven days these units could join up with the three Iraqi divisions already stationed near the Kuwaiti border (over 70,000 troops with more than 1,000 tanks), enabling Saddam Hussein once again to overrun Iraq's tiny neighbor.[78] While it is unclear what Saddam Hussein intended to do with the troop buildup, many feared that he was seeking to take advantage of a distracted United States, which was preoccupied with a military intervention in Haiti and a crisis on the Korean peninsula, and make gains against a largely defenseless Kuwait.[79] While there were U.S. combat aircraft in the theater—mainly in Saudi Arabia—to enforce the no-fly zone over southern Iraq, these aircraft were not configured to fly air-to-ground missions against armored forces. The only ground forces immediately available to try to fend off an Iraqi invasion were the outnumbered and overmatched four Kuwaiti brigades, which were moved to the border.[80] The USN had five major combatants in the Persian Gulf, along with the USS *Tripoli* ARG with 2,000 embarked Marines from the 15th Marine Expeditionary Unit (MEU), which was making a port call in UAE.[81] In short, U.S. forces were not postured to stop or even slow a heavy invasion force like the one that had massed in southern Iraq.

The situation quickly changed, however, as the Clinton administration rapidly reinforced the U.S. military presence in the region by deploying more than 25,000 troops. First, on Friday, October 7, the *George Washington* CVBG was ordered to leave its station in the Adriatic Sea, where it was supporting operations in Bosnia, and to move to the Red Sea so that it could support operations in Kuwait.[82] Around the same time, the Marines from the 15th MEU were loaded onto the ARG within 24 hours of being ordered to leave. They were in position off the coast of Kuwait in another 24 hours. As the MEU arrived on station, the CVBG entered the Red Sea and was ready to support operations in Kuwait. The United States also deployed its maritime prepositioning ships from the Indian Ocean and Guam, carrying equipment for ground forces,

---

[78] Herr, 1996, p. 26.

[79] White, 1999, p. 28.

[80] Chris Hedges, "Kuwait Troops Vow to Fight This Time," *New York Times*, October 11, 1994.

[81] The USN combatants were one cruiser, the USS *Leyte Gulf*; one destroyer, the USS *Hewitt*; and three frigates, the USS *Davis*, USS *Reid*, and USS *Hall*. Herr, 1996, p. 28.

[82] Herr, 1996, p. 28. The CVBG included the cruiser USS *San Jacinto*, the destroyer USS *John Barry*, and the auxiliary ship *Kalamazoo*.

and several U-2 and RC-135 intelligence surveillance and reconnaissance aircraft from the United States.[83]

On the evening of October 8, CONUS-based U.S. ground forces began to move as the lead elements of the 24th Infantry Division (Mechanized) left Fort Stewart, Georgia, and 180 soldiers to operate two Patriot batteries were deployed from Fort Polk, Louisiana. Both Army units were airlifted to the Persian Gulf and fell in on prepositioned equipment sets.[84] By Monday, October 10, there were 1,800 soldiers on the ground in Kuwait.[85] CONUS-based combat aircraft began to move after the ground troop movement.[86] The 23rd Composite Wing, which had been participating in a Red Flag exercise at Nellis Air Force Base, had to first return to its home station at Pope and then transit to the Saudi Arabia and then Al Jaber Air Base in Kuwait late on October 8. Additionally, 15 F-15s from the 1st Fighter Squadron deployed and were in position within 72 hours of being given the order to move.[87] The first combat aircraft arrived on October 10. By October 12, there were 36,000 U.S. troops in the Middle East, including F-15s, F-16s, A-10s, F-117s, F-111s, AWACs, and JSTARS, and DoD was prepared to deploy another 70,000, if necessary.[88] In the span of 8 days, the United States had deployed about 28,000 military personnel.[89] By October 12, the Iraqi government announced that its forces had been conducting routine exercises as the units began to board trains bound for Baghdad.[90] The rapidity of the U.S. military response seems to have impressed Hussein and compelled him not to test Washington's resolve. While Iraq continued to precipitate crises with its behavior, it never again massed forces in a menacing way against its neighbors.

Operation Deliberate Force

On August 29, 1995, a NATO air package consisting of 43 strike aircraft escorted by 14 SEAD (suppression of enemy air defenses) aircraft flew from Aviano Airbase and the *Theodore Roosevelt* carrier in the Adriatic to launch the first strikes into Bosnia as a part of Operation Deliberate Force.[91] The UN and NATO had been conducting military relief and peacekeeping operations in the Balkans since 1992 in an effort to minimize the humanitarian costs of the civil

---

[83] Herr, 1996, pp. 26–29.

[84] Rachel Schmidt, *Moving US Forces: Options for Strategic Mobility*, Washington. D.C.: Congressional Budget Office, January 1, 1997, p. 38.

[85] Herr, 1996, p. 29.

[86] DoD, Office of the Assistant Secretary of Defense (Public Affairs), "DoD News Briefing, Major General John Sheehan, Saturday October 8, 1994, 4:00 pm," October 8, 1994.

[87] DoD, Office of the Assistant Secretary of Defense (Public Affairs), October 20, 1994.

[88] White, 1999, p. 31.

[89] White, 1999, p. 32.

[90] White, 1999, p. 31.

[91] Mark J. Conversino, "Executing Deliberate Force, 30 August–14 September 1995," in Owen, 2000, p. 136.

wars that were ravaging the region. Operation Deliberate Force was the extension of these humanitarian operations that had gradually grown in scope over the preceding years. What began as a humanitarian airlift evolved into a UN peacekeeping operation with UN-defended safe zones on the ground and a no-fly zone enforced by NATO.[92] Because NATO aircraft were already patrolling the skies over Bosnia-Herzegovina when the operation began, NATO was able to be highly responsive and began strike operations the same day that the order was given.

The event that precipitated the NATO intervention was a Serb mortar attack against a marketplace in Sarajevo on August 28, which killed 38 civilians and wounded 80.[93] Yet there was no urgent threat to civilians or UN peacekeepers or any military rationale for swift action. Instead, the most pressing reason for action was to seize the moment of relative political unanimity to force a resolution of the Bosnian conflict before the consensus for action within the two international organizations collapsed.[94] In part, NATO and the UN had been galvanized to go on the offensive because of their failure to respond in July to the Serb offensive and ethnic cleansing in Srebrenica, a UN safe area. Although speed was an important factor, an equally (if not more important) consideration was the need for Operation Deliberate Force to succeed at minimal cost to NATO, the UN, and civilians.

With Operation Deliberate Force, the UN and NATO aimed to punish the Bosnian Serbs for violating the UN safe zones and to prevent further attacks against UN-designated safe areas.[95] Additionally, the operation sought to bring about a negotiated resolution to the Bosnian civil war by degrading the capability of the Bosnian Serb army, thereby forcing its leaders to sue for peace.[96]

Deliberate Force had been planned well in advance of the mortar attack that precipitated the operation. After a Serb SA-6 surface-to-air missile had shot down a USAF F-16C on June 2, Commander of Allied Air Forces Southern Europe, U.S. Lieutenant General Michael E. Ryan, had requested and received additional aircraft, especially jammers and HARM shooters.[97] In total, the number of aircraft participating in Operation Deny Flight—the no-fly zone over Bosnia—expanded by almost 20 percent in the summer of 1995.[98] At the same time, NATO had

---

[92] For more on the earlier operations, see Frederick J. Shaw Jr., "Crisis in Bosnia: Operation Provide Promise," in Timothy Warnock, ed., *Short of War: Major USAF Contingency Operations, 1947–1997*, Washington, D.C.: Air Force History and Museums Program, 2000, pp. 197–208; Daniel L. Haulman, "Resolution of the Bosnian Crisis: Operation Deny Flight," in Timothy Warnock, ed., *Short of War: Major USAF Contingency Operations, 1947–1997*, Washington, D.C.: Air Force History and Museums Program, 2000b, pp. 219–228.

[93] Mueller, 2000, pp. 20–21.

[94] Robert C. Owen, "Operation Deliberate Force, 1995," in John Andreas Olsen, ed., *A History of Air Warfare*, Washington, D.C.: Potomac Books, 2010, p. 202.

[95] Mueller, 2000, p. 28.

[96] Campbell, 2000, p. 87.

[97] Sargent, 2000, pp. 202–204.

[98] Sargent, 2000, p. 202.

planned a number of different offensive operations and been waiting for a Serb action that would serve as a trigger, allowing them to undertake one of them.[99]

At the outset, NATO had 280 aircraft from the United States (USAF, USN, and USMC), France, Britain, Turkey, the Netherlands, Italy, Spain, and Germany and NATO AWACs to conduct the air campaign.[100] Also when the operation began, NATO and the UN recalled the 61 "on-call aircraft," which soon joined the operation.[101] In short, there were an "abundance of forces available," to strike a "rather limited target set."[102] Consequently, some of Commander of Allied Air Forces Southern Europe's early requests for forces were canceled. One notable request that was not fulfilled was for F-117 Nighthawks. U.S. Secretary of Defense William Perry approved the deployment of 6 of the stealth fighters on September 9, but the Italian government, which was miffed at not being consulted in advance about whether it would provide base access, refused to allow the Nighthawks to be based at Aviano.[103] Given that the approved target list was nearly exhausted on September 7, the absence of the Nighthawks did not adversely affect the conduct of operations.[104] When Deliberate Force ended on September 20, there were 141 U.S. aircraft and a total of 350 alliance aircraft participating in the operation.[105]

While NATO had ample capacity to conduct strikes in Bosnia, its forces were often not permitted to do so because all offensive operations had to be authorized by both NATO and UN political and military leaders.[106] France and the United Kingdom, in particular, often refused to approve offensive operations that might put their UN Protection Force peacekeepers on the ground at risk.[107] This dual-key command-and-control construct necessitated a slow and laborious target selection process. Moreover, pilots had to abide by "some of the most restrictive [rules of engagement] ROEs in the history of air warfare," which in an effort to minimize collateral damage seriously constrained the types of targets that could be attacked and the amount of force that could be used.[108]

NATO forces were able to begin offensive operations the day that Deliberate Force was authorized because they had been planning for such an event for some time and had nearly all

---

[99] Owen, Robert C., ed., *Deliberate Force: A Case Study in Effective Air Campaigning*, Maxwell Air Force Base, Ala.: Air University Press, 2000, p. 213.

[100] Conversino, 2000, p. 133.

[101] Sargent, 2000, p. 202.

[102] Campbell, 2000, p. 112.

[103] Conversino, 2000, p. 151; Haulman, 2000b, p. 227.

[104] Campbell, 2000, p. 115.

[105] Sargent, 2000, p. 202; Conversino, 2000, p. 133.

[106] Owen, 2000, p. 206.

[107] Owen, 2000, p. 210.

[108] Owen, 2000, p. 214–215.

of the forces in place in the region and on alert. While the military situation did not hinge on a rapid response, the political situation did. Support for offensive operations within NATO and the UN was mixed, and many nations only agreed to begin the air war under the condition that it was kept under strict controls, which included a limited target list and strict rules of engagement. NATO was able to bring to bear enough power in the 17-day air campaign to convince Bosnian Serb leaders to comply with the demands placed on them, which ultimately led to the Dayton Accords.

Operation Allied Force

On March 24, 1999, NATO began OAF to end Serbian violence against ethnic Albanians in Kosovo and to prevent the conflict from spreading to other regions. OAF was expected to be a short coercive aerial campaign to achieve limited objectives, namely, compelling Yugoslav President Slobodan Milosevic to stop ethnically cleansing the Kosovar Albanians. Unlike many other operations examined, this was a war of choice that was initiated by NATO at the time and place of its choosing. In fact, NATO, and especially some member nations where the use of force was unpopular, clearly hoped that it would not have to follow through on its military threat and only resorted to force as a last resort. Responsiveness—in particular, demonstrating that NATO forces had the capability to follow through on the alliance's coercive threat—likely pushed Milosevic to the negotiating table in the fall of 1998, but ultimately this show of force proved unable to permanently stop the ethnic cleansing. Instead, it appears as if Milosevic did not doubt that NATO had the capability to launch a damaging air campaign, but he questioned whether the alliance had the political will to follow through on its coercive threat. Even if the alliance coalesced around the idea of an air campaign, Milosevic may have believed that this consensus would be ephemeral and that he could survive a war limited in intensity and attacks from one domain (the air).[109] Ultimately, Milosevic was wrong as NATO conducted a 78-day air war over Kosovo until he capitulated.

NATO initially threatened air strikes in the fall of 1998 when there was growing evidence of Serb atrocities against ethnic Albanians in Kosovo. On October 13, 1998, NAC, the political decisionmaking body for NATO, issued an ACTORD, which mobilized NATO forces and placed them under the control of the Supreme Allied Commander, General Wesley Clark. Consequently more than 400 aircraft, including 260 U.S. aircraft stationed in Europe, were alerted and prepared to launch strikes against the armed forces of Yugoslavia, which were controlled by Milosevic and responsible for most of the violence against Albanians.[110] Several days later Milosevic acceded to negotiations on Kosovar autonomy and permitted unarmed

---

[109] Posen, 2000, p. 51.

[110] Pierre Lhuillery, "Holbrooke Meets Milosevic as NATO Prepares for Possible Strikes," Agence France Press, October 11, 1998.

monitors from the Organization for Security Cooperation in Europe to verify compliance with the cease-fire.[111] In the fall of 1998, the threat to use force, coupled with steps to increase the readiness of NATO air forces and transfer them to NATO command, was sufficient to compel Milosevic to back down. But the truce did not hold for long.

NATO suspended plans for an air campaign against Yugoslav forces but kept the ACTORD in place, although offensive operations would require additional NAC approval.[112] It was only after a Serbian attack in the village of Racak on January 15, 1999, that resulted in the deaths of 45 Albanian citizens that NATO returned to planning for an air war. By January 21, over 200 combat aircraft had been deployed to Italy, and a U.S. aircraft carrier and eight other warships sailed to the Adriatic so that they were in position to launch attacks against Yugoslav forces, which were massing in Kosovo.[113] On January 30, NAC granted NATO Secretary General Javier Solana authority to order strikes in Yugoslavia.[114] As it increasingly appeared that NATO might employ force, the U.S. secretary of defense ordered additional aircraft, including 12 F-117 stealth fighters, 10 EA-6B Prowlers, 29 tanker aircraft, and 7 B-52s, to deploy to Europe. These U.S. reinforcements arrived within four days of being ordered to deploy, bringing the total NATO aircraft available for operations to 430.[115] Negotiations between the Kosovar Albanians and Milosevic continued until mid-March when the Yugoslav delegation left, refusing to sign an agreement. The next day, March 20, Serb forces began a major campaign to push ethnic Albanians out of their homes in Kosovo.[116] Three days later, Secretary General Solana ordered the Supreme Allied Commander Europe to begin air strikes.

OAF commenced on March 24, 1999, with 214 U.S. aircraft operating from European airbases, 2 B-2 bombers flying from CONUS, 130 allied aircraft, and 4 USN warships and several cruise-missile armed submarines in the AO.[117] NATO had planned for and its members were only willing to approve a graduated air campaign, which applied limited amounts of force in the hope of compelling Milosevic to capitulate, while minimizing the risk to NATO forces and civilians on the ground. Consequently, the number of aircraft in the theater that were prepared to undertake strike missions far exceeded the number of approved targets. For a target

---

[111] Lambeth, 2001, p. 7.

[112] DoD, 2000, p. A-4.

[113] "Department of Defense Regular Briefing: Briefer: Kenneth Bacon the Pentagon," Federal News Service, January 21, 1999.

[114] DoD, 2000, p. A-6.

[115] "Department of Defense Regular Briefing: Briefer: Captain Michael Doubleday the Pentagon," Federal News Service, February 18, 1999.

[116] DoD, 2000, p. A-7.

[117] DoD, 2000, p. 31.

to be approved, there was a convoluted and often lengthy vetting process that required the accession of all 19 members of the alliance.[118]

Nevertheless, General Clarke continued to request additional aircraft along with an Army attack aviation and fires unit in the hopes of gaining permission to expand operations against the Serb forces.[119] On April 3, the USS *Theodore Roosevelt* CVBG was ordered to turn around and return to the Mediterranean after initially being ordered to continue on to the Persian Gulf. It did not arrive on station for several more days, approximately two weeks into the operation.[120] By April 10, there were approximately 400 U.S. aircraft participating in the operation. Six days later, another 63 U.S. aircraft had arrived.[121] General Clark continued to request reinforcements, including as many as 300 additional aircraft. By the end of the campaign in early June, there were 720 U.S. aircraft in the theater.[122]

On June 9, 1999, after 78 days of NATO bombing operations, Milosevic capitulated. Despite the fact that it took much longer than anticipated, OAF is widely considered to have been a success; however, responsiveness was largely unimportant given the limited nature of the operation. Having ready and capable forces seems to have pushed Milosevic to the negotiating table in the fall of 1998, but the lack of political unanimity within NATO made him question whether NATO would follow through in the spring of 1999. In the end, U.S. forces were more than responsive enough to execute the OAF air plan, but the plan itself failed to quickly achieve the desired outcomes. It is not clear if more forces or a more intense bombing campaign would have yielded a different result faster. Over time, the fact that NATO remained unified, that additional air forces continued to flow into the theater, and that the list of approved targets was expanding to include strategic targets in Serbia seems to have pushed Milosevic toward negotiations.[123]

Operation Desert Fox

On December 16, 1998, the United States launched four days of air strikes against Iraq in Operation Desert Fox with the intent of punishing Iraqi President Saddam Hussein for his intransigence and reducing his ability to pose a threat to the region. Beginning in 1997, Iraqi president Saddam Hussein started to impede the efforts of UN weapons inspectors to monitor

---

[118] At the end of the first week, there were only 100 approved targets. Lambeth, 2001, p. 199.

[119] The deployment of the Army forces in Task Force Hawk was fraught with challenges. See Nardulli et al., 2002.

[120] Lambeth, 2001, p. 30.

[121] DoD, Office of the Assistant Secretary of Defense (Public Affairs), April 10, 1999; U.S. Department of Defense Office of the Assistant Secretary of Defense (Public Affairs), April 16, 1999.

[122] U.S. Department of Defense Office of the Assistant Secretary of Defense (Public Affairs), June 2, 1999.

[123] Posen, 2000, pp. 72–73. There were other factors at play as well, including Russian support. See Posen, 2000, pp. 66–75; Lambeth, 2001, pp. 67–86.

Iraqi weapons programs. This led to a crisis in early 1998, which was averted, after the United States. announced a large military buildup in the Persian Gulf region, inducing Hussein to sign to a new agreement that outlined procedures for the inspection of sensitive sites. Nevertheless, in August 1998, Hussein once again began to challenge UN inspectors and banned all inspections on October 31. The U.S. government insisted that Iraq must comply with UN Security Council resolutions, which obligated him to cooperate with the inspections.

As the crisis mounted, on November 11, the U.S. Secretary of Defense William Cohen ordered the deployment of additional forces to the Persian Gulf region, including 4,000 troops, 12 F-117s, an American expeditionary force composed of 6 B-1s and 36 fighters, and 12 B-52s to deploy to Diego Garcia.[124] The *Enterprise* Battle Group was also ordered to speed up its planned arrival to the Middle East. The following day, the B-52s departed for Diego Garcia along with additional tanker aircraft and the first detachment of airmen from the air and space expeditionary force (AEF).[125] While the other deployments were still ongoing, 7 B-52s had arrived at Diego Garcia and took off on a mission to launch cruise missile strikes at Iraq on the evening of November 14.[126] Several hours after taking off, the B-52 strikes were aborted as Hussein backed down and agreed to more inspections.

The remaining deployments were called off, with some aircraft, such as the F-117s, stopping in Europe before returning to their home stations. The Army halted the deployment of its forces from Georgia and the Patriot batteries.[127] It was not long before Hussein defied the UN inspectors again, prompting U.S. Central Command to begin planning for another attack on Iraq. This time, however, an emphasis was placed on maintaining tactical surprise to deny Hussein another opportunity to back down at the last minute. To avoid tipping off Hussein, the U.S. military planners decided that Operation Desert Fox would only employ forces already stationed in the theater.[128]

Operation Desert Fox involved four waves of nighttime strikes against Iraq, beginning with U.S. forces launching more than 200 sea-based cruise missiles at Iraq from ten warships and submarines in the AO on December 16. Additionally, naval and Marine Corps fighters from the USS *Enterprise* flew strikes against targets in southern Iraq. The sea-based fighters' range was limited because of a lack of land-based supporting aircraft, including tanker, command and control aircraft, and escorts.[129] This was followed by a second wave of 90 air-launched cruise

---

[124] White, 1999, p. 57.

[125] "200 Airmen from McGuire AFB Scheduled to Deploy to Gulf Region," Associated Press State and Local Wire, November 12, 1998.

[126] "Showdown with Iraq: UN Security Council Works Towards Consensus; Military Buildup Continues," *CNN Sunday Morning*, November 15, 1998.

[127] Cordesman, 1998, p. 108.

[128] Knights, 2005, pp. 199–200.

[129] Knights, 2005, pp. 200–202.

missiles fired from B-52 bomber aircraft based in Diego Garcia and British Tornado strikes on December 17.[130] The following evening, B-1 bombers and F-16s and British Tornados comprised the third wave of attacks, and the fourth and final night, there was a final wave of strikes that included F-15s, F-16s, and Tornados.[131]

Over the four-day operation, the United States launched 325 Tomahawk Land Attack Missiles and 90 air-launched cruise missile and dropped more than 600 bombs, hitting 98 targets, of which 43 were destroyed or serious damaged, 30 moderately damaged, 12 lightly damaged, and 13 not damaged.[132] Instead of these actions forcing Hussein to comply with sanctions, Iraqi defiance of the no-fly zone increased after Operation Desert Fox as challenges became a nearly daily occurrence.[133] If the aim of the operation, therefore, was to force Iraqi compliance, it failed. Responsiveness only mattered in the ability to conduct this operation insomuch as U.S. forces were in theater and therefore did not indicate that the United States was preparing for an operation and provide Hussein with time to back down and thwart them. In the long term, Desert Fox seems more successful in achieving its stated aim of limiting Hussein's ability to produce weapons of mass destruction, to degrade Iraqi command and control, and to overall reduce Iraq's ability to pose a threat to its neighbors. After the 2003 war, it became apparent that the limited 1998 strikes had done more than had been realized at the time to weaken the Hussein military.

---

[130] Knights, 2005, p. 203.

[131] Anthony H. Cordesman, *The Air Defense War After Desert Fox*, Washington, D.C.: Center for Strategic and International Studies, July 1, 1999, p. 2.

[132] Alfred B. Prados and Kenneth Katzman, "Iraq-U.S. Confrontation," Washington, D.C.: Congressional Research Service, November 20, 2001, p. 4.

[133] Knights, 2005, p. 62.

Appendix B: Operations Included in Simultaneity Analysis

Table B.1. Operations in Simultaneity Analysis

Operation Name	Start Date	End Date
Coup in Haiti	January 12, 1946	January 14, 1946
Signal U.S. Commitment to Turkey and Greece	March 22, 1946	April 5, 1946
Berlin Airlift	March 15, 1948	October 30, 1949
Arab-Israeli War	June 18, 1948	July 23, 1948
Korean War	June 25, 1950	July 27, 1953
Taiwan Patrol Force	June 27, 1950	January 1, 1979
Yugoslavia Crisis	March 15, 1951	November 1, 1951
Magic Carpet Haji Baba	August 24, 1952	August 29, 1952
Humanity Dutch Floods	February 2, 1953	February 17, 1953
Kyushu Flood	July 2, 1953	July 3, 1953
Foodlift (ROK)	July 27, 1953	August 1, 1953
Ionian Islands Earthquake	August 13, 1953	August 17, 1953
Dien Bien Phu	March 15, 1954	July 12, 1954
Guatemala-Honduras Crisis	May 20, 1954	June 29, 1954
PRC shoot down of Cathay Pacific	July 23, 1954	July 26, 1954
Operation Mercy	August 1, 1954	September 1, 1954
Passage to Freedom-Vietnam Evacuation	August 11, 1954	May 20, 1955
First Taiwan Strait Crisis	September 12, 1954	May 1, 1955
Salud	September 29, 1954	October 7, 1954
Honduran Elections	October 1, 1954	October 12, 1954
Tampico Flood	September 20, 1955	October 28, 1955
Typhoon Louise Iwo Jima	September 26, 1955	October 1, 1955
Snowbound	February 12, 1956	February 19, 1956
Suez nationalized	July 26, 1956	October 8, 1956
Cuban Civil War NEO	October 23, 1956	October 30, 1956
Suez War	October 29, 1956	December 21, 1956
Hungarian Refugees	November 4, 1956	December 1, 1956
Port Lyautey French-Moroccan Tensions	November 29, 1956	February 7, 1957
Safe Haven 1 and 2	December 11, 1956	June 30, 1957
Jordan Crisis	April 25, 1957	May 3, 1957
Locust Insecticide Tunisia	June 27, 1957	June 30, 1957
PRC-ROC tensions	July 1, 1957	September 30, 1957
Syrian Crisis	August 18, 1957	October 29, 1957

Operation Name	Start Date	End Date
Indonesian Unrest SCS Force	December 10, 1957	June 1, 1958
Abortive Coup Indonesia	February 21, 1958	May 20, 1958
Attach Nixon Motorcade Venezuela	May 13, 1958	May 15, 1958
Blue Bat	July 15, 1958	October 25, 1958
Second Taiwan Strait Crisis	August 28, 1958	December 18, 1958
Berlin Deadline	November 27, 1958	September 30, 1959
Laos Crisis	July 1, 1959	September 1, 1959
Typhoon Vera	September 26, 1959	February 1, 1960
Amigos Chilean Airlift	May 23, 1960	June 23, 1960
Agadir Earthquake	May 23, 1960	June 23, 1960
Sahara/New Tape	July 8, 1960	June 30, 1964
Guatemalan and Nicaraguan Unrest	November 14, 1960	December 1, 1960
1961 Laos Crisis	January 11, 1961	November 1, 1961
Bakwanga Famine Relief	January 26, 1961	February 9, 1961
Bay of Pigs	April 15, 1961	April 24, 1961
Dominican Republic Trujillo Assassinated	May 30, 1961	June 10, 1961
Kuwait Crisis	July 4, 1961	July 7, 1961
Exercise Big Truck	August 1, 1961	September 30, 1961
Berlin Crisis	September 5, 1961	July 1, 1962
Escalation in Vietnam	November 14, 1961	August 4, 1964
Dominican Crisis	November 18, 1961	December 19, 1961
Somali Flood Relief	November 18, 1961	November 18, 1961
North German Flood	February 18, 1962	February 20, 1962
Guatemala Unrest	March 1, 1962	April 1, 1962
Tanganyikan Flood Relief	April 25, 1962	June 6, 1962
Nam Tha Crisis	May 6, 1962	June 12, 1962
Guantanamo Harassment	July 25, 1962	July 27, 1962
Haiti Disorder	August 1, 1962	August 15, 1962
Iranian Disaster Assistance	September 3, 1962	November 12, 1962
Yemen Revolution	September 26, 1962	April 15, 1963
Cuban Missile Crisis	October 14, 1962	November 20, 1962
Long Skip	November 2, 1962	August 31, 1963
Typhoon Karen Guam	November 11, 1962	November 30, 1962
Clean Lens (Moroccan Flood)	January 9, 1963	January 15, 1963
Lifeline	April 1, 1963	September 1, 1963
Haitian Unrest	April 26, 1963	June 3, 1963
Blue Boy	July 27, 1963	August 8, 1963
Zanzibar Coup	January 12, 1964	January 13, 1964
Panama Canal	January 14, 1964	April 3, 1964

Operation Name	Start Date	End Date
Cyprus Peacekeepers	April 3, 1964	June 13, 1964
Guantanamo Harassment	May 1, 1964	May 7, 1964
Pakistani Floods	June 26, 1964	July 24, 1964
Vietnam War	August 5, 1964	March 29, 1973
Yugoslavia Flood	October 29, 1964	November 14, 1964
South Vietnam Flood	November 11, 1964	January 1, 1965
Dragon Rouge	November 19, 1964	December 2, 1964
Tunisian Bridge Collapse	November 25, 1964	November 29, 1964
Airlift for Danish UN PKO	March 30, 1965	May 23, 1965
Power Pack	April 27, 1965	September 21, 1965
Yemen War	July 1, 1965	August 30, 1965
Nice Way/Elder Blow	September 11, 1965	September 21, 1965
Indonesian Unrest	October 2, 1965	October 9, 1965
Project Refugee	October 19, 1965	November 25, 1965
Northern Italian Floods	November 11, 1966	November 12, 1966
1967 Six-Day War	June 6, 1967	June 11, 1967
Typhoon Sarah Wake Island	September 17, 1967	September 26, 1967
Pueblo Hijacking	January 23, 1968	December 23, 1968
Bombing of Cambodia	March 18, 1969	August 15, 1969
Search and Rescue of Shootdown EC-121	April 15, 1969	April 26, 1969
Cairo Agreement	October 22, 1969	November 3, 1969
Peruvian Earthquake	June 2, 1970	July 3, 1970
PFLP Hijacking	June 11, 1970	June 17, 1970
Black Sept Jordan Crisis	September 3, 1970	October 8, 1970
Typhoons Joan and Kate	October 19, 1970	October 27, 1970
Bonny Jack	June 17, 1971	July 17, 1971
Indo-Pakistani War	December 10, 1971	December 12, 1971
Bahama Lines	December 15, 1971	January 31, 1972
Operation Saklolo	July 21, 1972	August 15, 1972
F-4cs to Taiwan	November 1, 1972	November 1, 1974
Managua Earthquake	December 23, 1972	January 30, 1973
Nicaraguan	April 2, 1973	May 19, 1973
Project Scoot	April 11, 1973	October 1, 1973
Authentic Assistance	May 15, 1973	November 10, 1973
Pakistani Flood	August 20, 1973	September 22, 1973
Yom Kippur War	October 6, 1973	November 14, 1973
Night Reach	November 2, 1973	December 30, 1973
Nimbus Star/Nimbus Moon	April 10, 1974	December 20, 1974
King Grain	July 13, 1974	October 21, 1974

Operation Name	Start Date	End Date
Cyprus Coup	July 15, 1974	July 24, 1974
Cypriot Refugees	August 7, 1974	September 1, 1974
Hurricane Fifi	September 19, 1974	October 15, 1974
Cyprus Demonstrations	January 18, 1975	January 21, 1975
Vietnam Refugee Evacuations	April 4, 1975	September 16, 1975
Eagle Pull	April 12, 1975	April 12, 1975
Mayaguez Rescue	May 12, 1975	May 15, 1975
Angolan War of Independence	September 7, 1975	November 3, 1975
Jamaican Unrest	January 1, 1976	January 1, 1976
Polisario Rebels Morocco	January 5, 1976	January 22, 1976
Guatemala Earthquake	February 4, 1976	June 30, 1976
Beirut Evacuations	March 30, 1976	July 28, 1976
Typhoon Pamela Guam	May 23, 1976	June 9, 1976
Kenya-Uganda Tension	July 8, 1976	July 27, 1976
Korea Tree Incident	August 18, 1976	September 8, 1976
Turkish Earthquake	November 26, 1976	November 29, 1976
Turkish Earthquake 2	January 20, 1977	January 22, 1977
Shaba 1	March 1, 1977	May 30, 1977
Transport Americans out of Ethiopia	April 23, 1977	April 29, 1977
Zaire 1 and 2	May 16, 1978	June 16, 1978
Lift UN Fact Finding Mission to Namibia	August 8, 1978	August 8, 1978
Nicaragua	September 16, 1978	October 1, 1978
Guyanese Disaster	November 19, 1978	December 22, 1978
Iranian Revolution	December 8, 1978	February 17, 1979
Prize Eagle	January 1, 1979	January 30, 1979
PRC Invasion Vietnam	February 25, 1979	March 15, 1979
Flying Star	March 1, 1979	June 6, 1979
Yemen Conflict	March 6, 1979	June 6, 1979
Yugoslavian Earthquake	April 18, 1979	April 20, 1979
Evacuation of Nicaragua	June 12, 1979	August 31, 1979
Caribbean Storms	August 31, 1979	November 21, 1979
Soviet Troops Cuba	October 2, 1979	November 16, 1979
Afghanistan/Iran Hostage Crisis	October 9, 1979	January 21, 1981
Evacuation of Bolivia	November 7, 1979	November 7, 1979
Assassination of ROK President	November 26, 1979	December 26, 1979
Project Valentine Assistance	December 2, 1979	December 28, 1979
Election Monitoring in Zimbabwe	December 19, 1979	December 27, 1979
Lift UN Peacekeepers from Rhodesia	March 5, 1980	March 7, 1980
Eagle Claw	April 15, 1980	April 24, 1980

Operation Name	Start Date	End Date
Korea Unrest Post Kwangju	May 25, 1980	June 28, 1980
Vietnamese Incursion in Thailand	June 1, 1980	June 1, 1980
Elf One Iran-Iraq War	September 30, 1980	April 15, 1989
Algerian Earthquake Relief	October 12, 1980	October 23, 1980
Italian Earthquake	November 26, 1980	December 2, 1980
Creek Sentry	December 10, 1980	May 1, 1981
Airlift Salvadoran Government	January 1, 1981	January 1, 1981
Iran Hostage Release	January 20, 1981	January 25, 1981
Morocco Show of Force	January 29, 1981	February 7, 1981
Liberia	April 10, 1981	May 10, 1981
Syria-Israel Bekaa Crisis	May 3, 1981	September 14, 1981
Gambian Evacuation	August 8, 1981	August 8, 1981
Gulf of Sidra	August 12, 1981	September 1, 1981
AWACS after Sadat Assassinated	October 15, 1981	October 29, 1981
Response to DPRK Mobilization	December 1, 1981	December 1, 1981
Elf Sentry	March 19, 1982	December 31, 1982
Project Elsa	March 31, 1982	May 31, 1982
Israeli Invasion Lebanon	June 8, 1982	July 20, 1982
El Salvador Resupply	June 21, 1982	August 1, 1982
Chad Withdrawal	June 23, 1982	July 2, 1982
Chadian Food Airlift	July 6, 1982	July 14, 1982
Sinai PKO	August 6, 1982	September 5, 1982
PKO to Evacuate PLO from Lebanon	August 10, 1982	September 8, 1982
Lebanese Refugee Relief	August 23, 1982	August 24, 1982
Sabra and Shatila	September 22, 1982	February 11, 1983
Sudan-Libya Crisis	February 13, 1983	February 22, 1983
Burmese Invasion Supply Thailand	April 1, 1983	April 1, 1983
Operation BAT	May 1, 1983	May 1, 1983
Honduras Nicaragua	June 14, 1983	October 22, 1983
Senior Look	July 2, 1983	August 19, 1983
Assistance to Chad & Sudan	July 25, 1983	December 31, 1983
Korean Air Lines Flight 007	September 1, 1983	September 15, 1983
Rubber Wall	September 3, 1983	September 25, 1983
El Salvador Supply	October 1, 1983	October 1, 1983
Iran Threat to Block Oil	October 8, 1983	January 1, 1984
Beirut Bombings	October 23, 1983	December 9, 1983
Urgent Fury	October 24, 1983	November 3, 1983
Turkish Earthquake	November 1, 1983	November 5, 1983
U.S. Military Support Element Grenada	November 3, 1983	June 11, 1985

Operation Name	Start Date	End Date
Lebanon PKO	December 1, 1983	January 8, 1984
El Salvador Elections	February 1, 1984	July 31, 1984
Lebanon Withdrawal	February 21, 1984	April 26, 1984
Eagle Lift	March 19, 1984	April 9, 1984
Addition Deployments to KSA	June 1, 1984	June 1, 1984
Escort Ships	May 30, 1984	July 7, 1984
Monitoring Fighting in Chad	August 1, 1984	August 1, 1984
Chad Insurgency	August 1, 1984	September 1, 1984
Intense Look (Minesweeping the Red Sea)	August 3, 1984	October 2, 1984
Threats to Beirut Embassy	September 21, 1984	November 1, 1984
Gandhi	October 23, 1984	October 23, 1984
African Famine Relief	December 22, 1984	March 9, 1985
Lebanon NEO	March 1, 1985	March 15, 1985
Caribbean Drug Interdiction	April 5, 1985	April 20, 1985
TWA Hostage Release	June 14, 1985	July 24, 1985
Mexico Earthquake	September 19, 1985	September 30, 1985
Asian Games ISR	September 20, 1985	October 5, 1985
Achille Lauro Hijacking	October 7, 1985	October 12, 1985
Puerto Rican Mudslides	October 9, 1985	October 16, 1985
Egypt Air Hijacking	November 23, 1985	December 3, 1985
Yemen Civil War	January 1, 1985	January 31, 1986
Gulf of Sidra II	January 26, 1986	March 24, 1986
Afghan Refugees	March 1, 1986	July 1, 1993
El Dorado Canyon	April 14, 1986	April 15, 1986
Blast Furnace	July 1, 1986	November 15, 1986
Asian Games in South Korea	September 20, 1986	October 5, 1986
El Salvador Earthquake	October 10, 1986	November 7, 1986
USS *Stark*	May 26, 1987	May 26, 1987
Earnest Will	July 22, 1987	September 26, 1988
Haiti Unrest	January 1, 1988	January 31, 1988
Jittery Prop	January 8, 1988	December 14, 1988
Golden Pheasant	March 17, 1988	March 27, 1988
Nimrod Dancer	March 18, 1988	December 20, 1989
Issue Froth	April 1, 1988	April 20, 1988
Valiant Boom	April 5, 1988	April 11, 1988
Praying Mantis (Retaliation for Stark)	April 18, 1988	April 19, 1988
Support for Sudanese Refugees	May 1, 1988	August 11, 1988
Post Road	August 15, 1988	August 28, 1988
Burma Unrest	September 1, 1988	September 30, 1988

Operation Name	Start Date	End Date
Hurricane Gilbert	September 13, 1988	February 7, 1989
Korean Olympics	September 17, 1988	October 2, 1988
Strong Support (Hurricane Mitch)	October 1, 1988	November 1, 1988
Locusts Senegal	November 16, 1988	November 30, 1988
Armenian Earthquake	December 9, 1988	February 9, 1989
Lebanon Civil War	February 1, 1989	March 17, 1989
UN Airlift to Namibia	March 5, 1989	May 1, 1989
Africa 1	April 1, 1989	April 15, 1989
Panama Elections	May 11, 1989	June 29, 1989
Anchor Mark	August 24, 1989	September 5, 1989
Africa 2	September 29, 1989	October 15, 1989
Classic Resolve (Philippine Coup Attempt)	December 1, 1989	December 9, 1989
Operation Just Cause	December 15, 1989	January 3, 1989
Typhoon Ofa	February 6, 1990	March 1, 1990
Sharp Edge	June 2, 1990	January 9, 1991
Philippines Earthquake	July 17, 1990	August 1, 1990
Desert Shield	August 6, 1990	January 16, 1991
Kuwait Refugees	September 18, 1990	September 28, 1990
Somalia NEO	January 2, 1991	January 11, 1991
Desert Storm	January 17, 1991	March 1, 1991
Provide Comfort 1	April 5, 1991	July 15, 1991
Sea Angel	May 10, 1991	June 13, 1991
Ethiopian Drought	June 1, 1991	September 1, 1991
Fiery Vigil	June 8, 1991	July 2, 1991
Albanian Government Collapse	July 1, 1991	August 1, 1991
Provide Comfort 2/Northern Watch	July 15, 1991	March 17, 2003
Quick Lift	September 27, 1991	October 3, 1991
Angolan Civil War Relief	October 1, 1991	November 1, 1991
Victor Squared	October 2, 1991	November 11, 1991
Humanitarian Assistance to Ukraine	October 23, 1991	October 23, 1991
Humanitarian Supplies to the Former USSR	December 17, 1991	December 22, 1991
Promote Liberty	February 1, 1992	March 1, 1992
Provide Hope	February 1, 1992	September 1, 1994
Silver Anvil	May 3, 1992	May 4, 1992
Provide Promise	July 1, 1992	January 1, 1996
Search for Escobar	July 1, 1992	July 4, 1992
Sharp Guard	July 1, 1992	December 20, 1996
Maritime Monitor	July 16, 1992	November 22, 1992
Intrinsic Action	August 2, 1992	August 20, 1992

Operation Name	Start Date	End Date
Southern Watch	August 2, 1992	May 19, 2003
Provide Transition	August 12, 1992	October 7, 1992
Provide Relief/Restore Hope 1	August 14, 1992	February 28, 1993
Typhoon Omar	August 29, 1992	September 25, 1992
Impressive Lift	September 12, 1992	September 29, 1992
Typhoon Iniki	September 12, 1992	October 18, 1992
Sky Monitor	October 15, 1992	April 12, 1993
Silver Compass	October 23, 1992	October 25, 1992
Tajikistan Unrest	October 25, 1992	October 25, 1992
Strikes vs Iraqi Integrated Air Defense System	January 13, 1993	January 13, 1993
Restore Hope 2	March 1, 1993	March 27, 1994
ISR Ecuador	April 19, 1993	April 24, 1993
Deny Flight	April 12, 1993	August 29, 1995
UN Monitors to Cambodia	May 17, 1993	May 29, 1993
Tomahawk Land Attack Missile strikes Iraq	June 26, 1993	June 26, 1993
Able Sentry	July 5, 1993	July 12, 1993
Somalia Show Force	August 25, 1993	August 27, 1993
Shutdown 6	August 26, 1993	August 30, 1993
Support Democracy	September 1, 1993	October 18, 1993
Paraguay ISR	October 7, 1993	October 11, 1993
Distant Runner	April 9, 1994	April 16, 1994
Liberia NEO	May 1, 1994	May 1, 1994
Yemen NEO	May 7, 1994	May 9, 1994
Provide Assistance	May 12, 1994	July 21, 1994
Support Hope	July 22, 1994	September 1, 1994
Distant Haven	August 19, 1993	October 31, 1994
Uphold Democracy	September 8, 1994	April 17, 1995
Vigilant Warrior	October 8, 1994	December 15, 1994
Project Sapphire	November 21, 1994	November 23, 1994
United Shield	January 7, 1995	March 25, 1995
UN Airlift	February 3, 1995	February 10, 1995
Safe Border	March 1, 1995	October 24, 1998
DPRK Crisis	April 1, 1995	April 30, 1995
Quick Lift (Croatia)	June 30, 1995	August 10, 1995
Nomad Vigil	July 1, 1995	October 26, 1995
Medical Supplies to Belarus	July 23, 1995	July 23, 1995
Vigilant Sentinel	August 1, 1995	March 22, 1996
Deliberate Force	August 30, 1995	December 20, 1995
Joint Endeavor	December 20, 1995	February 28, 1999

Operation Name	Start Date	End Date
Sentinel Lifeguard	February 1, 1996	March 1, 1996
Taiwan Deterrent	March 1, 1996	April 17, 1996
Nomad Endeavor	March 1, 1996	February 1, 1997
Assured Response	April 8, 1996	August 3, 1996
Quick Response	May 20, 1996	June 22, 1996
Response to Khobar Towers	June 25, 1996	June 29, 1996
Burundi Civil War Evacuations	September 4, 1996	September 4, 1996
Desert Strike	September 3, 1996	September 4, 1996
Pacific Haven	September 15, 1996	September 19, 1996
Guardian Assistance	November 14, 1996	December 27, 1996
SFOR (Joint Guard, Joint Forge; NATO: Deliberate Guard, Deliberate Forge)	December 20, 1996	December 2, 2004
Assured Lift	February 18, 1997	March 3, 1997
Silver Wake	March 14, 1997	March 26, 1997
Guardian Retrieval	March 17, 1997	June 5, 1997
Noble Obelisk	May 27, 1997	June 5, 1997
Firm Response Passive Oversight	June 8, 1997	June 18, 1997
Passive Oversight	July 1, 1997	July 31, 1997
Bevel Edge	July 1, 1997	July 31, 1997
Desert Thunder	October 1, 1997	May 27, 1998
Indonesia Forest Fires	October 12, 1997	December 4, 1997
Silent Assurance	November 4, 1997	November 17, 1997
Phoenix Scorpion	November 19, 1997	November 25, 1997
Noble Response	January 21, 1998	March 25, 1998
Noble Safeguard	February 16, 1998	April 13, 1998
Allied Force	March 24, 1998	June 20, 1999
Bevil Incline	May 15, 1998	May 24, 1998
Safe Departure	June 6, 1998	June 17, 1998
Shepherd Venture	June 10, 1998	June 17, 1998
Determined Falcon	June 13, 1998	June 17, 1998
Balkan Calm	July 3, 1998	November 15, 1998
Determined Response	August 7, 1998	August 11, 1998
Autumn Shelter	August 9, 1998	August 11, 1998
Silver Knight	August 14, 1998	August 23, 1998
Sudan/Afghan Strikes	August 20, 1998	August 20, 1998
Shadow Express	September 24, 1998	October 13, 1998
Phoenix Duke	October 11, 1998	November 7, 1998
Eagle Eye	November 1, 1998	March 23, 1999
Desert Viper	November 4, 1998	November 19, 1998

Operation Name	Start Date	End Date
Shining Presence	December 10, 1998	January 6, 1999
Desert Fox	December 16, 1998	December 19, 1998
Christmas Island NEO	January 9, 1999	January 10, 1999
Task Force Falcon	June 12, 1999	March 18, 2014
Stabilise	September 20, 1999	February 28, 2000
Balkan Calm 2	October 16, 1999	November 18, 1999
Antarctica NEO	October 16, 1999	October 16, 1999
Determined Response	October 12, 2000	October 15, 2000

NOTES: ISR: intelligence, surveillance, and reconnaissance; NEO: noncombatant evacuation operation; PFLP: Popular Front for the Liberation of Palestine; PKO: peacekeeping operation; PLO: Palestine Liberation Organization; SCS: South China Sea; SFOR: Stabilisation Force in Bosnia and Herzegovina.

References

Appleman, Roy E., *South to the Naktong, North to the Yalu (June–November 1950)*, Washington, D.C.: Center of Military History, U.S. Army, 1992.

Armbrister, Trevor, *A Matter of Accountability: The True Story of the* Pueblo *Affair*, New York: Lyons, 2004.

Associated Press State and Local Wire, "200 Airmen from McGuire AFB Scheduled to Deploy to Gulf Region," November 12, 1998.

Barnett, Thomas, and Linda Lancaster, *Answering the 9-1-1 Call: U.S. Military and Naval Crisis Response Activity, 1977–1991*, Alexandria, Va.: Center for Naval Analysis, 1992.

Betts, Richard K., "Surprise Attack: NATO's Political Vulnerability," *International Security*, Vol. 5, No. 4, Spring 1981, pp. 117–149.

———, *Military Readiness: Concepts, Choices, Consequences*, Washington, D.C.: Brookings Institution, 1995.

Butler, Michael, "U.S. Military Intervention in Crisis, 1945–1994," *Journal of Conflict Resolution*, Vol. 47, No. 2, 2003, pp. 226–248.

Campbell, Christopher M., "The Deliberate Force Air Campaign Plan," in Owen, 2000, pp. 87–130.

Chivvis, Christopher S., *Toppling Qaddafi: Libya and the Limits of Liberal Intervention*, New York: Cambridge University Press, 2014.

Chun, Clayton K. S., *The Last Boarding Party: The USMC and the SS* Mayaguez *1975*, Oxford: Osprey, 2011.

Clausewitz, C. V., *On War*, M. Howard and P. Paret, ed. and trans., Princeton, N.J.: Princeton University Press, 1976.

CNN Sunday Morning, "Showdown with Iraq: UN Security Council Works Toward Consensus; Military Buildup Continues," November 12, 1998.

Cobble, W. Eugene, H. H. Gaffney, and Dmitry Gorenburg, *For the Record: All U.S. Forces' Responses to Situations, 1970–2000 (with Additions Covering 2000–2003)*, Alexandria, Va.: Center for Naval Analysis, June 2003.

Cohen, Eliot A., ed., *Gulf War Airpower Survey*, Vol. 5, Washington, D.C.: U.S. Department of Defense, 1993.

Conversino, Mark J., "Executing Deliberate Force, 30 August–14 September 1995," in Owen, 2000.

Cordesman, Anthony H., *Desert Fox: Key Official U.S. and British Statements and Press Conferences*, Washington, D.C.: Center for Strategic and International Studies, January 31, 1998.

———, *The Air Defense War After Desert Fox*, Washington, D.C.: Center for Strategic and International Studies, July 1, 1999.

Daalder, Ivo H., and Michael E. O'Hanlon, *Winning Ugly: NATO's War to Save Kosovo*, Washington, D.C.: Brookings Institution, 2000.

Davis, Richard G., *On Target: Organizing and Executing the Strategic Air Campaign Against Iraq*, Washington, D.C.: U.S. Air Force, 2002.

DoD—*See* U.S. Department of Defense

Dragnich, George S., *The Lebanon Operation of 1958: A Study of the Crisis Role of the Sixth Fleet*, Arlington, Va.: Center for Naval Analysis, September 1970.

Fairchild, Brian R., and Walter S. Poole. *The Joint Chiefs of Staff and National Policy, 1957–1960*, Vol. 7, Washington, D.C.: Office of Joint History, 2000.

Fehrenbach, T. R., *This Kind of War*, Washington, D.C.: Brassey's, 1963.

Field, James A., Jr., *History of United States Naval Operations: Korea*, Washington, D.C.: U.S. Department of the Navy, 1962.

Forster, Larissa, *Influence Without Boots on the Ground: Seaborne Crisis Response*, Newport, R.I.: Naval War College Press, 2010.

Futrell, Robert Frank, *The United States Air Force in Korea, 1950–1953*, Washington, D.C.: Office of Air Force History, 1983.

Gaffney, H. H., *Warning Time for U.S. Forces' Responses to Situations*, Alexandria, Va.: Center for Naval Analysis, June 2002.

Gaffney, Hank, Robert Benbow, Gregory Suess, and Glen Landry, *Employment of Amphibious MEUs in National Responses to Situations*, Alexandria, Va.: Center for Naval Analysis, 2006.

Garrett, John, *Task Force Smith: The Lesson Never Learned*, Fort Leavenworth, Kan.: School of Advanced Military Studies, U.S. Army Command and General Staff College, 2000.

Gough, Terrence J., *U.S. Army Mobilization and Logistics in the Korea War: A Research Approach*, Washington, D.C.: Center of Military History, U.S. Army, 1987.

Grant, Rebecca, "Airpower Made It Work," *Air Force Magazine*, November 1999. As of February 12, 2020:
https://www.airforcemag.com/article/1199airpower/

Halperin, M. H., *The 1958 Taiwan Straits Crisis: A Documented History*, Santa Monica, Calif.: RAND Corporation, RM-4900-ISA, 1966. As of February 12, 2020: https://www.rand.org/pubs/research_memoranda/RM4900.html

Harper, Jon, "New Crisis Response Force Gets Ready to Deploy to Middle East," *Stars and Stripes*, September 29, 2014. As of February 12, 2020: https://www.stripes.com/new-crisis-response-force-gets-ready-to-deploy-to-middle-east-1.305613

Harrison, Todd, "Rethinking Readiness," *Strategic Studies Quarterly*, Vol. 8, No. 3, Fall 2014, pp. 38–68.

———, "Trump's Bigger Military Won't Necessarily Make the US Stronger or Safer," *Defense One*, March 16, 2017. As of February 13, 2020: https://www.defenseone.com/ideas/2017/03/trumps-bigger-military-wont-necessarily-make-us-stronger-or-safer/136212/

Haulman, Daniel L., "Crisis in Southeast Asia: *Mayaguez* Rescue," in Timothy Warnock, ed., *Short of War: Major USAF Contingency Operations, 1947–1997*, Washington, D.C.: Air Force History and Museums Program, 2000a, pp. 105–114.

———, "Resolution of the Bosnian Crisis: Operation Deny Flight," in Timothy Warnock, ed., *Short of War: Major USAF Contingency Operations, 1947–1997*, Washington, D.C.: Air Force History and Museums Program, 2000b, pp. 219–228.

Hedges, Chris, "Kuwait Troops Vow to Fight This Time," *New York Times*, October 11, 1994.

Henriksen, Dag, "Inflexible Response: Diplomacy, Airpower and the Kosovo Crisis, 1998–1999," *Journal of Strategic Studies*, Vol. 31, No. 6, 2008, pp. 846–850.

Herbert, Adam J., "The Balkan Air War," *Air Force Magazine*, March 2009. As of February 13, 2020: https://www.airforcemag.com/article/0309balkan/

Herr, W. Eric, *Operation Vigilant Warrior: Conventional Deterrence Theory, Doctrine, and Practice*, Maxwell Air Force Base, Ala.: School of Advanced Airpower Studies, June 1996.

Keaney, Thomas A., and Eliot A. Cohen, *Revolution in Warfare? Air Power in the Persian Gulf*, Annapolis, Md.: Naval Institute, 1995.

Kiraly, Emery M., Robert C. Owen, and Aron Pinker, *Part II: Chronology of the Gulf War*, in Eliot A. Cohen, ed., *Gulf War Airpower Survey*, Vol. 5, Washington, D.C.: U.S. Department of Defense, 1993.

Knights, Michael, *Cradle of Conflict: Iraq and the Birth of the Modern U.S. Military*, Annapolis, Md.: Naval Institute, 2005.

Lambeth, Benjamin S., *NATO's Air War for Kosovo: A Strategic and Operational Assessment*, Santa Monica, Calif.: RAND, MR-1365-AF, 2001. As of February 13, 2020: https://www.rand.org/pubs/monograph_reports/MR1365.html

———, "The U.S. Is Squandering Its Airpower," *Washington Post*, March 5, 2015.

Leffler, Melvyn P., *A Preponderance of Power: National Security, the Truman Administration, and the Cold War*, Stanford, Calif.: Stanford University Press, 1994.

Lerner, Mitchell B., *The* Pueblo *Incident: A Spy Ship and the Failure of American Foreign Policy*, Lawrence: University Press of Kansas, 2002.

Lhuillery, Pierre, "Holbrooke Meets Milosevic as NATO Prepares for Possible Strikes," Agence France Press, October 11, 1998.

Little, Robert D., and Wilhelmine Burch, *Air Operations in the Lebanon Crisis of 1958*, USAF Historical Division Liaison Office, Washington, D.C., October 1962, declassified on February 23, 1982.

Mahoney, Robert, *U.S. Navy Responses to International Incidents and Crises, 1955–1975: Survey of Navy Crisis Operations*, Alexandria, Va.: Center for Naval Analysis, 1977.

Marolda, Edward J., *The United States Navy and the Persian Gulf War*, Naval History and Heritage Command (website), August 23, 2017. As of February 13, 2020: https://www.history.navy.mil/research/library/online-reading-room/title-list-alphabetically/ u/the-united-states-navy-and-the-persian-gulf.html

Matthews, James K., and Cora J. Holt, *So Many, So Much, So Far, So Fast: United States Transportation Command and Strategic Deployment for Operation Desert Shield/Desert Storm*, Washington, D.C.: Joint History Office, 1992.

Mattis, James N., "Remarks by Secretary Mattis on the National Defense Strategy," U.S. Department of Defense (website), January 19, 2018. As of February 13, 2020: https://dod.defense.gov/News/Transcripts/Transcript-View/Article/1420042/remarks-by -secretary-mattis-on-the-national-defense-strategy

McCullough, Amy, "Cutting Readiness," *Air Force Magazine*, April 2013. As of February 13, 2020: https://www.airforcemag.com/article/0413readiness/

Mellinger, Philip S., *Airpower Myths and Facts*, Maxwell Air Force Base, Ala.: Air University Press, December 2003.

Millett, Alan R., and Peter Maslowski, *For the Common Defense: A Military History of the United States of America*, New York: Free Press, 1994.

Moroney, Jennifer D. P., Stephanie Pezard, Laurel E. Miller, Jeffrey G. Engstrom, and Abby Doll, *Lessons from Department of Defense Disaster Relief Efforts in the Asia-Pacific Region*, Santa Monica, Calif.: RAND, RR-146-OSD, 2013. As of February 13, 2020: https://www.rand.org/pubs/research_reports/RR146.html

Mueller, Karl P., "The Demise of Yugoslavia and the Destruction of Bosnia: Strategic Causes, Effects, and Responses," in Owen, 2000, pp. 1–36.

———, "Flexible Power Projection for a Dynamic World: Exploiting the Potential of Air Power," in Cindy Williams, ed., *Holding the Line: US Defense Alternatives for the Early 21st Century*, Cambridge, Mass.: MIT Press, 2001.

———, ed., *Precision and Purpose: Airpower in the Libyan Civil War*, Santa Monica, Calif.: RAND, RR-676-AF, 2015. As of February 13, 2020: https://www.rand.org/pubs/research_reports/RR676.html

Nardulli, Bruce R., Walter L. Perry, Bruce R. Pirnie, John Gordon IV, and John G. McGinn, *Disjointed War: Military Operations in Kosovo, 1999*, Santa Monica, Calif.: RAND, MR-1406-A, 2002. As of February 13, 2020: https://www.rand.org/pubs/monograph_reports/MR1406.html

Owen, Robert C., ed., *Deliberate Force: A Case Study in Effective Air Campaigning*, Maxwell Air Force Base, Ala.: Air University Press, 2000.

Owen, Robert C. "Operation Deliberate Force, 1995," in John Andreas Olsen, ed., *A History of Air Warfare*, Washington, D.C.: Potomac Books, 2010.

Patchin, Kenneth L., *Flight to Israel*, Scott Air Force Base, Ill.: Military Airlift Command, April 30, 1974.

Peters, John E., Stuart Johnson, Nora Bensahel, Timothy Liston, and Traci Williams, *European Contributions to Operation Allied Force*, Santa Monica, Calif.: RAND Corporation, MR-1391-AF, 2001. As of November 3, 2020: https://www.rand.org/pubs/monograph_reports/MR1391.html

Petraeus, David, and Michael O'Hanlon, "The Myth of a U.S. Military 'Readiness' Crisis," *Wall Street Journal*, August 9, 2016.

Pettyjohn, Stacie L., and Jennifer Kavanagh, *Access Granted: Political Challenges to the U.S. Overseas Military Presence, 1945–2014*, Santa Monica, Calif.: RAND, RR-1339-AF, 2016. As of February 13, 2020: https://www.rand.org/pubs/research_reports/RR1339.html

Poole, Walter S., *The Joint Chiefs of Staff and National Policy, 1973–1976*, Washington, D.C.: Office of Joint History, 2015.

Posen, Barry R., "The War for Kosovo: Serbia's Political-Military Strategy," *International Security*, Vol. 24, No. 4, Spring 2000, pp. 39–84.

Prados, Alfred B., and Kenneth Katzman, "Iraq-U.S. Confrontation," Washington, D.C.: Congressional Research Service, November 20, 2001.

Project AIR FORCE Assessment of Operation Desert Shield: The Buildup of Combat Power, Santa Monica, Calif.: RAND Corporation, MR-356-AF, 1994. As of February 11, 2020: https://www.rand.org/pubs/monograph_reports/MR356.html

Quandt, William B., *Soviet Policy in the October 1973 War*, Santa Monica, Calif.: RAND Corporation, R-1864-ISA, May 1976. As of February 13, 2020: https://www.rand.org/pubs/reports/R1864.html

———, *Peace Process: American Diplomacy and the Arab-Israeli Conflict Since 1967*, 3rd ed., Washington, D.C.: Brookings Institution, 2005.

Reed, Ronald M., "Chariots of Fire: Rules of Engagement in Operation Deliberate Force," in Owen, 2000, pp. 381–430.

Roberts, Ross, "Desert Fox: The Third Night," *Proceedings*, Vol. 125/4/1, April 1999. As of February 13, 2020: https://www.usni.org/magazines/proceedings/1999/april/desert-fox-third-night

Rotterman, Gordon, *Korean War Order of Battle: United States, United Nations, and Communist Ground, Naval, and Air Forces, 1950–1953*, Westport, Conn.: Praeger, 2002.

Rumbaugh, Russell, *Defining Readiness: Background and Issues for Congress*, Washington, D.C.: Congressional Research Service, June 14, 2017.

Sanborn, James K., "Deal with Spain Gives Marines Permanent Base, Surge Capability," *Marine Corps Times*, June 19, 2015. As of February 13, 2020: https://www.marinecorpstimes.com/news/your-marine-corps/2015/06/19/deal-with-spain -gives-marines-permanent-base-surge-capability/

Sargent, Richard L., "Aircraft Used in Deliberate Force," in Owen, 2000, pp. 199–256.

Scales, Robert H., *Certain Victory: The US Army in the Gulf War*, Fort Leavenworth, Kan.: U.S. Army Command and General Staff College Press, 1993.

Schmidt, Rachel, *Moving US Forces: Options for Strategic Mobility*, Washington. D.C.: Congressional Budget Office, January 1, 1997.

Schwartz, General Norton, "Air Force Contributions to Our Military and Our Nation," speech delivered at the World Affairs Council of Delaware, Wilmington, Del., January 20, 2012.

Shaw, Frederick J., Jr., "Crisis in Bosnia: Operation Provide Promise," in Timothy Warnock, ed., *Short of War: Major USAF Contingency Operations, 1947–1997*, Washington, D.C.: Air Force History and Museums Program, 2000.

Shlapak, David A., *The Russian Challenge*, Santa Monica, Calif.: RAND, PE-250-A, 2018. As of February 13, 2020:
https://www.rand.org/pubs/perspectives/PE250.html

Siegel, Adam B., *The Use of Naval Forces in the Post-War Era: U.S. Navy and U.S. Marine Corps Crisis Response Activity, 1946–1990*, Alexandria, Va.: Center for Naval Analysis, 1991.

———, *A Chronology of U.S. Marine Corps Humanitarian Assistance and Peace Operations*, Alexandria, Va.: Center for Naval Analysis, September 1994.

Stein, Janice Gross, "Deterrence and Compellence in the Gulf, 1990–1991: A Failed or Impossible Task," *International Security*, Vol. 17, No. 2, Fall 1992, pp. 147–179.

Stephens, Alan, "The Air War in Korea, 1950–1953," in John Andreas Olsen, ed., *A History of Air Warfare*, Washington, D.C.: Potomac Books, 2010.

Tan, Michelle, "Army Quick-Response Forces Stood Up Around the World," *Air Force Times*, November 10, 2013.

Tefteller, William R., *Strategic Airlift Support for U.S. Forces Deployment to Operation Desert Shield*, Washington, D.C.: National Defense University, 1991.

Tzu, Sun, *The Art of Warfare*, Lionel Giles, trans., New York: Race Point, 2017.

U.S. Air Force, *America's Air Force: A Call to the Future*, Washington, D.C., July 2014.

U.S. Congressional Budget Office, *Moving U.S. Forces: Options for Strategic Mobility*, Washington, D.C., February 1997.

U.S. Department of Defense, *Conduct of the Persian Gulf War: Final Report to Congress*, Washington, D.C.: Office of the Secretary of Defense, April 1992.

———, *Report to Congress, Kosovo/Operation Allied Force After Action Report*, Washington, D.C., January 31, 2000.

———, *Summary of the 2018 National Defense Strategy of the United States of America: Sharpening the U.S. Military's Competitive Edge*, Washington, D.C., 2017.

U.S. Department of Defense, Office of the Assistant Secretary of Defense (Public Affairs), "DoD News Briefing, Major General John Sheehan, Saturday October 8, 1994, 4:00 pm," October 8, 1994.

———, "Background Briefing: Subject Iraq," Washington, D.C., October 20, 1994.

———, "DoD News Briefing, Captain Michael Doubleday, February 18, 1999," February 18, 1999.

———, "DoD News Briefing, Saturday April 10, 1999 2:05 pm," April 10, 1999.

———, "DoD News Briefing, Friday April 16, 1999," April 16, 1999.

———, "DoD News Briefing, Wednesday June 2, 1999 2:10 pm," June 2, 1999.

U.S. Government Accountability Office, *Military Personnel: DoD Needs to Reevaluate Fighter Pilot Workforce Requirements*, Washington, D.C., April 11, 2018.

Van Staaveren, Jacob, *Air Operations in the Taiwan Crisis of 1958*, Washington, D.C.: USAF Historical Division Liaison Office, November 1962.

Vick, Alan J., Richard M. Moore, Bruce R. Pirnie, and John Stillion, *Aerospace Operations Against Elusive Ground Targets*, Santa Monica Calif.: RAND Corporation, MR-1398-AF, 2001. As of November 3, 2020:
https://www.rand.org/pubs/monograph_reports/MR1398.html

Warnock, A. Timothy, ed. *The U.S. Air Force's First War: Korea 1950–1953 Significant Events*, Montgomery, Ala.: Air Force History and Museums Program, Air Force Historical Research Agency, 2000.

Welsh, General Mark A., III, *Department of the Air Force Presentation to the Committee on Armed Services House of Representatives: Impact of Sequestration*, February 13, 2013. As of February 13, 2020:
https://www.armed-services.senate.gov/imo/media/doc/Welsh%2002-12-13.pdf

———, *Department of the Air Force Presentation to the Senate Armed Services Committee: The Impact of Sequestration on National Defense*, January 28, 2014. As of February 13, 2020:
https://www.armed-services.senate.gov/imo/media/doc/Welch_01-28-15.pdf

Wetterhahn, Ralph, *The Last Battle: The* Mayaguez *Incident and the End of the Vietnam War*, New York: Plume Group, 2002.

White, Lt Col Paul K., *Crises After the Storm: An Appraisal of U.S. Air Operations in Iraq Since the Persian Gulf War*, Washington, D.C.: Washington Institute for Near East Policy, 1999.

Wols, Donald K., "Relationship Between Cohesion and Casualty Rates: 1st Marine Division and 7th Infantry Division at Inchon and the Chosin Reservoir," thesis, Fort Leavenworth, Kan.: U.S. Army Command and General Staff College, 2004. As of February 13, 2020:
https://apps.dtic.mil/dtic/tr/fulltext/u2/a429076.pdf

Yaqub, Salim, *Containing Arab Nationalism: The Eisenhower Doctrine and the Middle East*, Chapel Hill: University of North Carolina Press, 2004.

Zenko, Micha, "Saving Lives with Speed: Using Rapidly Deployable Forces for Genocide Prevention," *Defense and Security Analysis*, Vol. 20, No. 1, March 2004, pp. 3–19.